CLINICAL CASE REPORTING IN EVIDENCE-BASED MEDICINE

MILOS JENICEK MD

Professor (McMaster University)
Professor Emeritus (Université de Montréal)
Adjunct Professor (McGill University)

Hamilton (Ontario) and Montreal (Quebec)
Canada

OXFORD   AUCKLAND   BOSTON   JOHANNESBURG   MELBOURNE   NEW DELHI

Butterworth-Heinemann
Linacre House, Jordan Hill, Oxford OX2 8DP
225 Wildwood Avenue, Woburn, MA 01801-2041
A division of Reed Educational and Professional Publishing Ltd

℞ A member of the Reed Elsevier plc group

First published 1999

British Library Cataloguing in Publication Data
Jenicek, Milos
  Clinical case reporting in evidence-based medicine
  1. Clinical medicine   2.   Evidence-based medicine – Case studies
  I. Title
  616

Library of Congress Cataloguing in Publication Data
Jenicek, Milos, 1935-
    [Casuistique médicale. English]
    Clinical case reporting in evidence-based medicine/Milos Jenicek.
      p.   cm.
    Includes bibliographical references and index.
    ISBN 0 7506 4592 X
    1. Medicine – Case studies.   2.   Communication in medicine.
  I. Title.
  RC66.J4613
  616'.09–dc21

ISBN 0 7506 4592 X

Printed and bound in Great Britain by Biddles Ltd, Guildford and Kings Lynn

CONTENTS

'. . . Any patient that goes
through
the door of this
hospital
is a potential case
report . . .'

Anonymous
(The Montreal General Hospital)

PREFACE

With the rise of strong clinical research in this century, case reports have been treated with ambivalence. On the one hand, serious researchers know that case reports fall far short in terms of scientific credibility. They cannot match the decisiveness of randomized controlled trials, the power of large cohort studies, or the elegance of case control studies in the right hands. On the other hand, physicians in practice like them. Clinicians prefer to learn from individual cases, either their own or a colleague's, and published case reports retain a strong connection with their simpler cousins, the case descriptions presented at the bedside or discussed over lunch.

In this book, Milos Jenicek helps us understand the place of case reports in medicine, especially the *'special cases that advance the knowledge, research and practice of medicine'*. He writes mainly about the cases that are described to a broad audience in journal articles and not about case reporting to local colleagues at the bedside. He recognizes that case reports are a large and neglected part of the medical literature and helps to raise their level of quality.

Case reports have long been familiar elements of medical journals and they remain so. During the period 1946–1976, single case reports comprised 13 per cent of articles in leading medical journals, and 38 per cent of articles were of ten or fewer patients and thus, for scientific purposes, were like single case reports. And case reports, in one form or another, are still a regular feature of major clinical journals today, albeit in various forms. *The Lancet* publishes case reports under just that heading in nearly every issue and takes great pains to make the cases interesting and instructive, even though most are not so rare as to be first reports. Other journals, such as *The New England Journal of Medicine* and *Annals of Internal Medicine,* have taken to publishing full research articles in their Brief Reports sections which were formerly reserved for case reports, but continue to publish accounts of single cases as letters – and sometimes even as original articles. *The New England Journal of Medicine* also uses individual cases to teach clinical medicine through its series Clinical Problem Solving. The less widely circulated peer reviewed journals, which comprise more of the world's medical journals, continue to publish single case reports, as they always have.

Can the enduring popularity of case reports be justified on scientific grounds ? After all, they can offer no credible evidence on the rate of clinical events, a central element in studies of the natural history of disease. And, because they include no comparison (control) group, they are severely limited as a way of understanding risk factors and causes of disease, and the effectiveness of therapeutic or preventive interventions.

But case reports do serve other purposes, also important to scientific progress. What can they do that large, controlled studies cannot?

Case reports are the only source of information on rare events. If one has a patient in the office with an unusual symptom – perhaps a potential drug reaction – where else can one turn for guidance, however limited ? Case reports are also a rich source of hypotheses. All that we know about clinical HIV infection or toxic shock syndrome began with observations of individual cases. Case reports can be a vehicle for teaching clinical medicine. For example, many physicians may have forgotten what they learned in medical school about patent ductus arteriosus and find in a properly presented case an opportunity to review cardiac physiology, even if the case itself is not unique. And yes, case reports can entertain. The former editor of the *British Medical Journal,* Steven Lock, pointed out that *'scientific journals have several functions: to inform, instruct, comment and, possibly, amuse'* in order to maintain a strong place in the lives of busy clinicians. For this reason, when my wife and I were editors of *Annals of Internal Medicine* we decided to publish a case report of a man and his dog, both stricken with histoplasmosis while he was chopping a rotten log, complete with chest radiographs of both patients.

Case reports are most valuable as part of a comprehensive database of other case reports, which can be searched when needed. Finding the case report one wants when one needs it is far more feasible now that Medline searching is free to users and the World Wide Web has made literature searching widely available throughout the world. After all, few clinicians read any one journal, and by definition they are unlikely to find in it information about the specific rare event that they need to know about.

Milos Jenicek has written a comprehensive book about case reports: what makes them useful, how they fit into the totality of scientific evidence, and how they should be described in writing. A few other scholars have written chapters and articles about case reports, describing their components and style, and what editors and reviewers looks for in the ones they publish. Dr Jenicek's book includes these elements in greater depth, but has also gone a step further. He places case reports in the context of modern clinical science by establishing connections between single case reports, case series and more elaborate and powerful studies, such as cohort and case control studies and clinical trials.

So case reports are not on the fringe of science and clinical practice, as they are sometimes believed to be. They deserve serious, scholarly consideration. Dr Jenicek has given them the attention they deserve.

ROBERT H. FLETCHER
Professor
Harvard Medical School
and Harvard School of Public Health

FOREWORD

William Osler's famous statement: *'The best teaching of medicine is that taught by the patient himself'* should be inscribed on the front door of any institution whose mission it is to train young physicians.

Medical diagnosis is the complex art of recognizing illnesses whose signs and symptoms constantly differ from case to case. This task is made all the more difficult by the patient's subjective description of his or her state of health. Thus, the physician's approach should be based on a meticulously organized analysis inevitably put to the test by the theoretical aspects of the pathology. In addition to this quasi-automated process, the clinician's thoughts should also progress through a series of intuitions, each providing hypotheses to be improved upon or rejected as determined by objective confirmatory elements, most often paraclinical in nature.

This phase of diagnosis, which is related to the theory of forms (Gestalt) used in sensory psychology, is entirely reliant on **clinical experience**, the acquisition of which is extremely time-consuming not only during training but also throughout the physician's professional life.

Clinical experience is in fact based on the **observation of clinical cases**.

All medical training, or at least that in the French tradition, relies on the relationship between three actors: the patient, the student and the teacher. The student is required to write down his observations pertaining to a case. The teacher then examines them in detail and presents a logical solution. Thus, attitudes are inherently learnt, pathological examples are committed to memory, and a reference tool much more effective than a paragraph in a book is obtained to aid in the preparation of diagnostic hypotheses.

However, there are times when a more unusual, deceptive case should be described to a larger group: members of a team in a hospital, participants at a scientific symposium, or even the entire medical profession nationally or internationally by means of a specialized journal.

In all these cases, the goal is the same: to convey a message that interests the audience and which is useful to everyone including future patients of an identical pathology.

This presentation of cases with the aim of focusing the medical profession's attention on the peculiar characteristics of an illness is an art that certain clinicians truly master.

In reality, the art of presenting cases should follow the rules set out in this book. Yet, it is also important to note from the start that there is a continuum between writing the details (in a medical chart) of the examination, diagnosis and the therapeutic follow-up that physicians must provide for each patient, followed by the preparation of a 'case report' which can be submitted to the review comittee of a prestigious journal.

In addition to its pedagogical benefits, the presentation of clinical cases has heuristic implications. A case marks the beginning of a series of cases and therefore an epidemiological study. It is also the starting point of physiopathological hypotheses leading to confirmatory experimental studies. Numerous examples of this exist and communicable pathologies are a painful reality that have lead to the opening of new nosological chapters stemming from a single case (e.g. AIDS, mad cow disease). Many other examples can also be found in pharmacology (the discovery of sulfonylureas for example), to the point where drug monitoring based on case studies has become a fully fledged discipline. In my area of specialty, endocrinology, it is well known that the fundamental role of the thyroid on all bodily functions was suggested to Kocher by the single observation of a patient who had recently undergone a thyroidectomy.

It is logical that Milos Jenicek, who is interested in teaching clinical research methodology, seeks to help our young colleagues improve their 'clinical case report' writing. Thus, Chapter 3 of this book, 'How to prepare a single case report', is an invaluable tool. This chapter covers everything, from the reasons for a clinical case report, the selection of data and the format of the discussion, to the choice of bibliographical material. It also gives semantic hints.

Milos Jenicek is of the North American school, and more specifically of the Canadian school, which prizes the theorization of medical activity. The notion of clinical epidemiology derives from these approaches and Milos Jenicek has produced a number of works on the topic. In one of these, he tackled a new domain of clinimetrics, a term coined by A.R. Feinstein. In this book, with regards to the relationship between clinical cases, he focuses on 'medical casuistics'. By way of a very clever semiotic analysis, he justifies the transfer of this theological term to the medical field.

Certainly, French medical culture is less avid for new concepts, rating scales and indexes than its North American equivalent, but since the author's background is varied, partly due to numerous teaching assignments in Europe, he seamlessly links together these two approaches.

Therefore, this work, as the author had hoped, *'helps young clinicians communicate clinical observations in a clear, comprehensible and complete manner'*. It is my belief that the result will be even more far-reaching than that. By 'coding' the process, Milos Jenicek provides a successful formula for rigorous analysis and presentation. He stresses, in fact, that *'no matter what the format of the presentation, the beginner will quickly come to realize that the evaluation of a case is actually the first task to accomplish'*.

Thus, we return to clinical teaching. By formalizing information recording, the reasoning process is supported. To paraphrase Boileau, *'that which is well designed can be clearly stated'*, and it is not by chance that the writing of a book, thesis or review has become a step in the thought pattern of the student. In addition, the emphasis placed on the quality of medical record keeping today will gradually lead to the inclusion of this parameter in the evaluation of physicians' activities.

The training of future evaluators according to a reading scale for documents that are presented to them is a final practical and positive result of this book.

RENÉ MORNEX
Professor Emeritus
Claude Bernard University
Corresponding member of the
French Academy of Medicine

Introductory comments: objectives and contents of this book

As long as medicine exists, physicians will share their clinical case experiences with their colleagues. However, if the unique, the exceptional and the extraordinary are not properly recorded, a good portion of the progress achieved in terms of patient care, originality, innovation, and enrichment of medical thinking will be lost, possibly forever.

The objective of this brief guide is to help young clinicians understand and prepare meaningful clinical case reports.

Clinicians have always recounted to their peers the circumstances, patient outcomes and consequences of the cases they have encountered. This is and will always be a powerful means of learning and further developing the practice of medicine.

Clinical cases are presented everywhere: on hospital floors, on daily ward rounds and on grand rounds, as well as in seminars, in meetings, in the medical press and at conventions. Wherever reports of this type are made, the initial physician–patient interaction marks the beginning of the acquisition of medical knowledge and of the professional and scientific retelling of lived experience.

A process of such importance cannot rely solely on gut feelings and on the natural intelligence of current and forthcoming generations. Unfortunately, our medical forefathers often played this trick on us. In reality, the form and content or 'direction' of case reporting must be learnt like anything else. And herein lies the purpose of this book.

The reader will be presented with six chapters, as outlined below.

In the first chapter, the concept and objectives of modern medical casuistics, i.e. the recording and study of cases of disease[1], will be explained.

In the second, the reader will see that medical casuistics or clinical case reporting is simply a part of a wider field of case studies in human sciences and a current end-product of a long history of thinking in the health, social and business sciences. In fact, clinical case reporting can be considered a component of the rapidly developing qualitative research field, and therefore complements its basic fundamental clinical, bedside clinical and field epidemiological quantitative constituents.

In the third chapter, step by step instructions cover the preparation of a good clinical case report, mainly for the purpose of publication in the medical press.

In the fourth, basic principles, methods and techniques of clinical case reporting are further illustrated using an annotated example of a good clinical case report as published recently in a medical journal. This chapter will demonstrate that an increasing number of presentations already effectively follow the guidelines set forth in this book.

The fifth chapter covers case series reports – the description and analysis of more than one case.

The sixth chapter summarizes the major points of the previous chapters and offers strategic tips for future clinical practice and research.

> **How this book should be read:**
> - If you are a busy intern or resident wanting to master only the essential steps of preparing a clinical case report, read Chapter 3.
> - If you are looking for a concrete example of a well-prepared published case, read Chapter 4.
> - If you want to report several cases as a case series report, read Chapter 5 (after reading Chapter 3).
> - If you are a clinical teacher or an intellectual luminary who feels it necessary to possess a broad knowledge of everything, read this book from cover to cover!

Many readers might ask why a clinical epidemiologist would tackle the area of clinical case reporting. There are two reasons for this. The first is that accurate descriptions of clinical cases now rely heavily on clinical epidemiology and clinimetrics (measurement and classification of clinical observations). The second is that, at the moment, when index cases of some new or otherwise relevant clinical phenomena are observed, elucidated and described, new hypotheses arise and further causal, treatment or prognostic research ideas are triggered.

We have learnt a great deal about the methodology required to adequately describe disease occurrence, to illustrate its cases, to demonstrate the most effective, efficient and efficacious treatment, to understand disease prognosis, to produce the best research synthesis and to make complex decisions in practice. However, we have not yet invested a similar effort in the area of clinical case reporting, the building block of all of these elements.

There are two levels of clinical case reporting. Routine cases are reported in day to day ward or office practice. This is of special interest to those in clinical service. Special cases that advance the knowledge, research and practice of medicine are also reported. This book focuses mainly on the latter.

Presently, clinical case reporting appears to be another necessary training tool for our graduate students in clinical epidemiology.

Is this subject entirely new? Not exactly. It is in some way a product of the ebb and flow of medicine between the Old and New Worlds. In the past century, many North American, Asian and African physicians travelled to Germany and France to master the thinking and workings of medicine at that time. They met such people as Pierre-Charles Alexandre Louis (1787–1872) who helped transform medicine into a science based on observation by rigorously following up on and systematically recording the vital functions of his patients. He also compiled other clinical observations, thus unknowingly becoming one of the forefathers of clinical epidemiology. Pinel, Laënnec, Bichat, Corvisart and Martinet were among those who personified the French school of medicine that focused on the detailed recording of history, disease manifestations and autopsies. Soon after, the German and Austrian schools (led by Mueller, Rokitanski and Skoda, among others) served as training grounds for North American physicians. Thus, William Osler was able to reiterate the importance of history and clinical examinations as a fundamental part of case recording and reports[2]. Generations have now elapsed since these migrations of ideas and experiences between the Old and the New Worlds.

Today, **evidence-based medicine**[3], born from Canadian experience and innovation, reverses the tide of medical reasoning and thinking and subjects it to the test of time. Since evidence-based medicine makes use of the most complete information available on diagnostic methods, disease occurrence and effectiveness of prevention or treatment, complete evidence from clinical cases must be provided.

It is important to realize that, before evidence of aetiology or treatment effectiveness is established, evidence of cases and their occurrence is required. Is this possible without a proper clinical case reporting methodology? Definitely not. Very occasionally, methodologically remarkable proofs of aetiology or treatment effectiveness are based only on 'mirages' of cases; they should not be. **Cases themselves are and must be the first line of evidence**.

The reader will benefit from an *a priori* knowledge of clinical epidemiology and clinimetrics. Although this monograph explains a few of the basic concepts of these fields, it is not a true substitute for *ad hoc* textbooks entirely devoted to them. This book does, however, strive to be accessible subsequent or parallel reading for the novice, and an essential tool for the busy clinician.

As William Makepeace Thackeray once said, *'the two most engaging powers of an author are to make new things familiar and familiar things new'*. This text tries to do both. It is based largely on my last work in French[4], which was perhaps the first to focus on today's modern medical casuistics. However, recent advances in clinical sciences compelled me to offer this more complete and updated message.

A small army of highly qualified and experienced friends and colleagues have reviewed and greatly improved my work: Carol-Ann Oldbury from the Montérégie Public Health Directorate (word processing and general text review), Jacques Cadieux of the Université de Montréal (infographics), Nicole Kinney of Linguamax (literary and stylistic review), Drs Karl Weiss and Denise Ouellette of the Maisonneuve-Rosemont Hospital (medicine and surgery) and Ann McKibbon, Cindy Walker-Dilks and Angela Eady of the McMaster University Health Information Research Unit (general readability review). All these people made this book possible and I am grateful to them.

REFERENCES

1. *Miller-Keane Encyclopedia & Dictionary of Medicine, Nursing & Allied Health*, 6th Edition. Philadelphia: WB Saunders, 1997.
2. Walker HK. The origins of the history and physical examination, pp. 22–28 in: *Clinical Methods. The History, Physical, and Laboratory Examination*, 3rd Edition. Edited by HK Walker, WD Wall and JW Hurst. Boston: Butterworths, 1988.
3. Sackett DL, Richardson WS, Rosenberg W, Haynes B. *Evidence-Based Medicine. How to Practice and Teach EBM.* New York: Churchill Livingstone, 1997.
4. Jenicek M. *Casuistique Médicale. Bien Présenter un Cas Clinique. (Medical Casuistics. Presenting a Clinical Case Well.)* St Hyacinthe and Paris: Edisem and Maloine, 1997.

Cover illustration:

André Brouillet, *Une leçon clinique à la Salpêtrière*, 1887. From the Collection des Hospices Civils de Lyon. Reproduced with permission.

Historical note (compliments of the Hospices Civils de Lyon):

'. . . At the 1887 Exposition, this painting drew crowds. Imposing in stature, colourful and spectacular, it depicted Dr Charcot lecturing on hypnosis in a quasi-religious manner before an appreciative somewhat non-medical audience. The painting was centred on a woman well known to the Salpêtrière (saltpetre factory), semi-reclining in the arms of Dr Brabinski as a result of the hypnosis.

Already a celebrated medal winner, chances were good that André Brouillet's work would impress the directors of the Beaux-Arts School, the largest provider of art for the public collections.

In those days, the State and cities owned a large number of works of art which they exhibited in newly created museums.

Following a common practice in the Third Republic, Brouillet asked the directors of the Beaux-Arts School to allow the State to buy the painting. Given the subject, its timeliness and its popularity, the request was almost unnecessary. The State purchased the painting for 3000 fr. It was assigned to the Musée de Nice by decree on 18th August 1891. Many years later, it was discovered in the archives of this museum by Professor Jean Lépine, a retired neurologist, who, after serving as Dean of the Faculty of Medicine in Lyon, retired and became the assistant mayor of Nice. Since then, the painting's role has become more prominent. In 1965, it was awarded by decree to the Pierre Wertheimer neurological hospital in Lyon. Presently being restored, it will hang in the committee room of the hospital. . .'

CHAPTER 1

The importance of modern case reporting or medical casuistics

CHAPTER 1

The importance of modern case reporting or medical casuistics

For many, a clinical report focused on an isolated case means nothing. In reality, if our peers do not select a relevant topic and present it properly, the medical community barely stifles a yawn.

1.1 CASE REPORTING AS A FIELD OF DISSENT

Recently, a British psychiatry resident complained in a Letter to the Editor of a journal[1]: '*after reading the case report "Suspicion of somatoform disorder in undiagnosed tabes dorsalis[2]" I found myself puzzled as to what I should learn from it. Tabes dorsalis is adequately described in most of the standard textbooks,*

---

and the fact that a psychiatric assessment was solicited before the investigation had confirmed the diagnosis, seems a curious reason for a case report, particularly such a long one. Good case reports are instructive and illuminating. Could I make a plea that, in view of the burgeoning numbers of case reports, only those which present truly novel observations be selected for publication?'. (NB The Editor defended the novelty of the observation.)

Was this case report really poorly presented? Was the resident at fault for not learning what he was supposed to? Or was it a little of both?

1.2 CLINICAL CASE REPORTING TODAY

This example shows that clinical case reporting today requires methodologies and selection criteria probably as stringent as that of any other basic or decision-oriented clinical research field.

Since antiquity, physicians have learnt from their more experienced peers, as well as from their own successes and failures. Accurate recounting of clinical experience continues to be essential to the progress of medicine, because it marks the beginning of the longer and increasingly complex process of learning initiated by observations made at the patient's bedside. For example, before the clinical characteristics and epidemiology of the Ebola virus related haemorrhagic fever were established, the first cases had to be correctly described. The analysis and subsequent synthesis of this new problem only occurred once this was done.

An interesting case, such as an unusual cerebral embolism or mycotic infection may become an 'index case', thus launching a search for similar occurrences and helping to determine the importance of frequency in similar cases within communities of interest. These steps may lead to the formulation of a hypothesis concerning the new diagnostic entity, its possible causes and treatment.

Nowadays, medicine is widening its focus and objectives, i.e. to prevent and treat disease, to protect individuals and communities, and to promote the best health possible in as many people as possible. It has replaced its paradigm of art and deterministic science with a paradigm of probabilistic science based on decisions made in uncertain situations.

Basic clinical or laboratory research, which is essentially explanatory, now has a counterpart in decision- and action-oriented 'bedside' clinical research. Clinical epidemiology and biostatistics have quickly become pivotal in this evolution of medicine where a new and better equilibrium has been established for art and science. At the end of such a learning process, the results obtained from observation and experience are reapplied at the starting point, namely, the clinical work for an individual patient and the care provided to communities.

If 'serious' medicine aspires to have its own 'letters of nobility', it should not be forgotten that everything begins with the personal experience of the physician and his patient, at the office or hospital. If this groundwork is not correct, the study of disease occurrence, aetiological research and clinical trials are worth very little. Similarly, medical intervention is of little value if there is poor handling of individual cases of the disease in question.

A resident on night call has admitted a new patient. He performs a complete work-up, gives his 'impression' of the major clues leading to the differential diagnosis and orders a treatment plan to 'stabilize' the patient. The next morning, the resident has to present this new case to his or her peers. If particularly interesting, the 'case' may be presented on floor rounds or grand rounds or ultimately reported in the medical press. However, floor or grand presentations and clinical case reporting in the medical press must abide by the specific rules detailed in the following chapters.

There are three levels of case reporting, each with different goals.

1. **Floor or daily ward presentations** often include all cases indiscriminately. These types of reports serve administrative purposes. They also ensure continuity and completeness of care.
2. Only selected cases are presented on **rounds.** The cases may be scientifically relevant but the reports themselves generally strive to ensure better operation of in-house work.
3. **Clinical case reports in the medical press** represent (or should represent) a scientific endeavour comparable to other observational or experimental research projects.

This book focuses mainly on the principles, methods and techniques of special clinical case presentations produced for the medical press. Compared to the more clearly established standards of in-house presentations, written clinical case reporting requires the mastery of additional rules (as detailed in Chapter 3) leading to clear, effective, comprehensible and complete presentations. Any intern or resident should follow these rules.

Yet we tend to derive our attitudes, knowledge, skills and experience from various sets of observations based on the current standards of clinical epidemiology. We know what to do with repeated observations and experience. However, do we know equally well how to study an 'individual' and how to share such an experience with others? Do we need more specific training in this area?

Responsible clinical researchers manage to publish two or three original papers in a year. In contrast, practising clinicians must, almost immediately, discuss their experiences with individual patients, either orally or in writing. If these practising clinicians prepare their reports correctly, and this book aims to help them achieve this goal, their contributions should be even more significant to the advancement of our medical wisdom.

Well-selected and adequately presented clinical cases are an important tool in the acquisition and comprehension of new information. Reports on the first cases of toxic shock syndrome or necrotizing fasciitis led to the study of series of such cases, and ultimately to the understanding of their causes and treatment.

Clinical case reporting in the medical press remains a controversial topic. However, if the report is good, its publication should be encouraged without hesitation[3–5].

Not so long ago, journals as distinguished as *Cancer* often simply discarded clinical case reports. Such decisions may have been motivated by the desire to include only preferred topics that the Editors had deemed 'absolute priorities'. These decisions might also have been made to avoid the ever-increasing flood of sometimes questionable information. How then is particularly bad reporting to be controlled?

Chew[6] recently analysed the fate of clinical case reports submitted to the *American Journal of Roentgenology*. Only one in five (20 per cent) reports submitted was effectively published.

However, it should be noted that there are many other reputed journals whose Editors encourage the publication of well-selected, relevant and impeccably presented cases. Such articles are an integral part of the medical culture since they enrich professional experience and lead to better clinical reasoning, often triggering 'more serious' research.

Case Records of the Massachusetts General Hospital in *The New England Journal of Medicine* or Clinical Reports in *The Lancet*, or Case Studies in *The Canadian Medical Association Journal*, or even issues of the *Journal of Obstetrics and Gynecology* devoted to special clinical cases, are just a few illustrations of these editorial policies.

1.3 CONTEMPORARY DEVELOPMENTS IN CLINICAL CASE REPORTING

The modern clinical case report was made possible by the recent development of clinical epidemiology[7–12], and of clinimetrics[9,13,14] in particular.

Any study of multiple cases must derive from the study of one individual case. The study of both individual and multiple cases requires adequate gathering of clinical data, hardening of soft data, conversion of data into clinical information, appropriate reproduction of the pathway from clinical observation to diagnosis, well-structured descriptions of the natural history and of the natural and clinical course of the disease, clear pictures of the gradient and spectrum of the pathology under study, and the selection of eloquent paraclinical data.

A statement highlighting the pragmatic questions to be answered by a case study and the operational results expected from such an endeavour has quickly become a requirement.

At present, we are inspired by the development and the results of research in human biology and pathology, by decision-oriented clinical research based on clinical epidemiology and biostatistics, and by the results of an evidence-based approach in medicine.

Individual case studies must catch up with other fields of medical research in their observational and analytical methodology, as well as in their interpretation and presentation.

The clinical case paradigm has become more defined and structured. Clinical epidemiology provides it with the tools necessary to make clinical case experience both operational and beneficial for the patient, and useful for the physician.

Today, general practitioners and specialists in training acquire an increasingly adequate clinical attitude, knowledge and set of skills. They also learn how to better read, interpret and integrate into their practice the results of original studies and papers[15–30]. Unfortunately, they do not always know how to prepare and present a clinical case[31] and how to retain the most important and useful information for their future practice.

In fact, we have searched in vain in the literature for a monograph devoted to clinical case reporting in terms of medical casuistics.

Hospital rounds often focus on the basics of the pathology, diagnosis and treatment[32] of relevant cases. However, case presentations are more involved than that. They represent an exercise in the organization of thought, in the acquisition of relevant clinical and paraclinical information, and in the making of correct clinical decisions.

Petrusa and Weiss[33] suggest that every physician in training should prepare and present at least one acceptable clinical case in terms of form and content.

The design and presentation of a clinical case report are not only important teaching tools. Supported by clinimetrics, they lead the authors of the reports towards other fields of clinical research. The following chapters outline the ideas and elements that constitute a structured approach to the work-up and presentation of clinical cases.

Are these details necessary? Definitely. Until now, only two textbooks on epidemiology have briefly touched upon the topic of desirable contents of case reports[34,35]. Other contributions to the field focus more on structure and style than on what should be in a report and what should be left out. None of these sources provides indications concerning the preparation, content and architecture of a case report for publication. Chapter 3 will meet this objective.

Despite the humble background of clinical case presentations, reporting unique experiences in comparison with aetiologic research or clinical trials will always remain a powerful aid in generating research and clinical experience.

REFERENCES

1. Pearson R. Case report criticism. (Correspondence) *Br J Psychiatry*,1992;**160**:280.
2. Fichtner CG, Weddington WW. Suspicion of somatoform disorder in undiagnosed tabes dorsalis. *Br J Psychiatry*,1991;**159**:573–5.
3. Soffer A. Case reports in the Archives of Internal Medicine. *Arch Intern Med*,1976;**136**:1090.
4. Nahum AM. The clinical case report: "Pot boiler" or scientific literature? *Head & Neck Surg*,1979;**1**:291–2.
5. Simpson RJ Jr, Griggs TR. Case reports and medical progress. *Persp Biol Med*,1985;**28**:402–6.
6. Chew FS. Fate of manuscripts rejected for publication in the AJR. *AJR*,1991;**156**:627–32.
7. Fletcher RH, Fletcher SW, Wagner EH. *Clinical Epidemiology – The Essentials*. Baltimore and London: Williams and Wilkins, 1982 (3rd Edition: 1996).
8. Jenicek M, Cléroux R. *Epidémiologie. Principes, Techniques, Applications. (Epidemiology. Principles, Techniques, Applications.)* St Hyacinthe and Paris: Edisem and Maloine,1982 (see Chapter 14).
9. Jenicek M, Cléroux R. *Epidémiologie Clinique. Clinimétrie. (Clinical Epidemiology. Clinimetrics.)* St Hyacinthe and Paris: Edisem and Maloine, 1985.
10. Feinstein AR. *Clinical Epidemiology. The Architecture of Clinical Research*. Philadelphia: WB Saunders,1985.
11. Sackett DL, Haynes RB, Tugwell PX. *Clinical Epidemiology: A Basic Science for Clinical Practice*. Boston: Little, Brown, 1984.
12. Jenicek M. *Epidemiology. The Logic of Modern Medicine*. Montreal: Epimed International, 1995.
13. Feinstein AR. An additional basic science for clinical medicine: IV. The development of clinimetrics. *Ann Intern Med*,1983;**99**:843–8.
14. Feinstein AR. *Clinimetrics*. New Haven: Yale University Press, 1987.
15. Gehlbach SH, Bobula JA, Dickinson JC. Teaching residents to read medical literature. *J Med Educ*,1980;**55**:362–5.
16. Oxman AD, Sackett DL, Guyatt GH for the Evidence-Based Medicine Working Group. Users' guides to the medical literature. I. How to get started. *JAMA*,1993;**270**:2093–5.
17. Guyatt GH, Sackett DL, Cook DJ for the Evidence-Based Medicine Working Group. Users' guides to the medical literature. II. How to use an article about therapy or prevention: A. Are the results valid? *JAMA*,1993;**270**:2598–601; B. What were the results and will they help me in caring for my patients? *JAMA*,1994;**271**:59–63.
18. Jeaschke R, Guyatt G, Sackett DL for the Evidence-Based Medicine Working Group. Users' guides to the medical literature. III. How to use an article about a diagnostic test: A. Are the results of the study valid? *JAMA*,1994;**271**:389–91; B. What are the results and will they help me in caring for my patients? *JAMA*,1994;**271**:703–7.

19. Levine M, Walter S, Lee H, Haines T, Holbrook A, Moyer V for the Evidence-Based Medicine Working Group. Users' guides to the medical literature. IV. How to use an article about harm. *JAMA*,1994;**271**:1615–9.

20. Laupacis A, Wells G, Richardson S, Tugwell P for the Evidence-Based Medicine Working Group. Users' guides to the medical literature. V. How to use an article about prognosis. *JAMA*,1994;**272**:234–7.

21. Oxman AD, Cook DJ, Guyatt GH for the Evidence-Based Medicine Working Group. Users' guides to the medical literature. VI. How to use an overview. *JAMA*,1994;**272**:1367–71.

22. Richardson SW, Destsky AS for the Evidence-Based Medicine Working Group. User's guides to the medical literature. VII. How to use a clinical decision analysis: A. Are the results of the study valid? *JAMA*,1995;**273**;1292–5; B. What are the results and will they help me in caring for my patients? *JAMA*,1995;**273**:1610–3.

23. Hayward RSA, Wilson MC, Tunis SR, Bass EB, Guyatt G for the Evidence-Based Medicine Working Group. Users' guides to the medical literature. VIII. How to use clinical practice guidelines: A. Are the recommendations valid? *JAMA*,1995;**274**:570–4.

24. Wilson MC, Hayward SA, Tunis SR, Bass EB, Guyatt G for the Evidence-Based Medicine Working Group. Users' guides to the medical literature. VIII. How to use clinical practice guidelines: B. What are the recommendations and will they help you in caring for your patients? *JAMA*,1995;**274**:1630–2.

25. Guyatt GH, Sackett DL, Sinclair JC, Hayward R, Cook DJ, Cook RJ for the Evidence-Based Medicine Working Group. Users' guides to the medical literature. IX. A method for grading health care recommendations. *JAMA*, 1995;**274**:1800–4.

26. Naylor CD, Guyatt GH, for the Evidence-Based Medicine Working Group. Users' guides to the medical literature. X. How to use an article reporting variations in the outcomes of health services. *JAMA*,1996;**275**:554–8.

27. Naylor CD, Guyatt GH for the Evidence-Based Medicine Working Group. Users' guides to the medical literature. XI. How to use an article about a clinical utilization review. *JAMA*,1996;**275**:1435–9.

28. Guyatt GH, Naylor CD, Juniper E, Heyland DK, Jaeschke R, Cook DJ for the Evidence-Based Medicine Working Group. Users' guides to the medical literature. XII. How to use articles about health-related quality of life. *JAMA*,1997;**277**:1232–7.

29. Drummond MF, Richardson SW, O'Brien BJ, Levine M, Heyland D for the Evidence-Based Medicine Working Group. Users' guides to the medical literature. XIII. How to use an article on economic analysis of clinical practice: A. Are the results of the study valid? *JAMA*,1997;**277**:1552–7.

30. O'Brien BJ, Heyland D, Richardson WS, Levine M, Drummond MF for the Evidence-Based Medicine Working Group. Users' guides to the medical literature. XIII. How to use an article on economic analysis of clinical practice: B. What are the results and will they help me in caring for my patients? *JAMA*,1997;**277**;1802–6.

31. Huston P, Squires BP. Case reports: Information for authors and peer reviewers. *CMAJ*,1996;**154**:43–4.
32. Loschen EL. The resident conference: A method to enhance academic intensity. *J Med Educ*,1980;**55**:209–10.
33. Petrusa ER, Weiss GB. Writing case reports: An educationally valuable experience for house officers. *J Med Educ*,1982;**57**:415–7.
34. Fletcher RH, Fletcher SW and Wagner EH. Case reports (Chapter 10), pp. 208–211 in: *Clinical Epidemiology. The Essentials*, 3rd Edition. Baltimore: Williams & Wilkins, 1996.
35. Jenicek M. General requirements of good case studies (Chapter 5, Section 5.2.4), pp. 135–136 in: *Epidemiology. The Logic of Modern Medicine*. Montreal: Epimed International, 1995.

Case studies, casuistics and casuistry in human sciences and medical culture

CHAPTER 2

Case studies, casuistics, and casuistry in human sciences and medical culture

In popular speech and thought, casuistics erroneously means an excessive subtlety and detailing. In contrast, medical casuistics is a process which allows the essential to be extracted from a given patient case or clinical situation.

2.1 WHAT IS CASUISTICS?

In medicine, case reporting plays only a limited role in the acquisition of new knowledge. However, its role is very important because it leads to more advanced research. In fact, medical practice often relies on this kind of presentation.

Today, **casuistics** signifies *'the recording and study of the cases of any disease'* [1]. More precisely, we can define it as an *'observation, analysis and interpretation of clinical cases'* [2]. A **casuist**, then, is a *'practitioner of the study of clinical cases'* [2].

The term 'casuistics' originates from the Latin word *casus* meaning an occurrence, a reality. In lexicography, casuistics refers to the application of general laws or rules to a particular subject or fact. *In general,* casuistics denotes the practice of solving problems by performing certain specific actions based on the general principles and study of similar cases[3]. However, casuistics *as a research method* focuses on the study of special individual cases from which a general rule of action can be derived. Medical casuistics is best defined by the latter explanation.

Several conflicting views of casuistics stem from the field of **casuistry**. In French culture and language, the very same word (*casuistique*) signifies **casuistics, casuistry** and **case studies** (*vide infra*). In English, all these entities are different[4–6].

Over time, casuistry has acquired different meanings in theology, philosophy, economics, administration and business, as well as in popular speech.

2.1.1 Historical comments – casuistry in philosophy, theology and ethics

In opposition to casuistics, **casuistry** is defined as *'a system of rules for distinguishing right from wrong in everyday situations, usually associated with a concept of morality that sees right conduct in terms of obedience to a set of closely defined laws'*. Casuistry was used by the Roman Stoics, Chinese Confucians, Jewish compilers of the Talmud, Muslim commentators of the

Qur'an, scholastic philosophers of Medieval Europe and later by Roman Catholic theologians. The fine distinctions employed by some Jesuit casuists compelled their opponents to equate casuistry with specious reasoning [7].

For theologians, casuistry is the part of Christian morality that focuses on states of conscience[8]. A casuist is a theologian who studies morality and offers his or her opinion to help solve cases or states of conscience[9].

Casuistry is also seen as a set of rules that distinguishes right from wrong in daily situations. Such a system is associated with a certain concept of morality that recognizes appropriate behaviour within the context of adherence to a well-defined set of laws. For example, the Talmud in Jewish faith, culture and tradition can be considered an old exercise in casuistry. In contrast, Quebec's Civil Code or similar codes in other parts of the world can be thought of as a modern example in casuistry.

From this historical point of view, health care philosophers or ethicists sometimes identify themselves as *'modern casuists'* [10,11] experienced in the philosophy and theology of morality. This should not be confused with our above-mentioned definition. Health care ethicists assist clinicians in their decisions by applying general arguments of morality to particular individual cases. Should a heart transplant be performed on an elderly patient suffering from another incurable disease? Who should have access to haemodialysis in situations when such access to this technology is limited?

Health care ethicists define casuistry as *'the art of applying abstract principles, paradigms and analogies to particular cases'* [12]. General rules and maxims are not universal and immovable, because they hold true only in relation to the typical situations of the agent and to the circumstances of the action[10].

The direction of inference in medical casuistics is quite different from the direction of inference in theology, philosophy, ethics or business (see Figure 2.1). While the latter apply general concepts to specific cases, medicine, in its will of

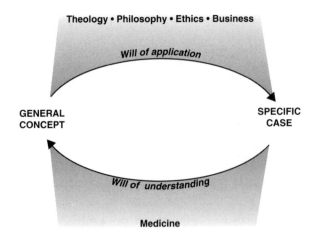

Figure 2.1 – Direction of Inference in Casuistics

understanding, starts by studying specific cases and seeks to use them as a basis for general concepts and as the starting point for further research of the problem illustrated by the cases.

A 'case' may be centred not only on an individual patient but also on a particular situation or event. Jonsen and Toulmin[13] describe the controversy surrounding a statement made by Geraldine Ferraro, the first woman to be an American vice-presidential candidate. Although Ms Ferraro was personally opposed to abortion, she declared that the decision to abort should be made by the expectant mother herself. An important discussion followed on what should prevail: general or universal principles or the particular situation of each and every individual under consideration. It was suggested that an equilibrium between the general and the specific be sought to avoid the challenge of generally accepted values. Because, in this case, the term 'casuistry' could be considered pejorative, its replacement by the French term 'casuistics' (in French: 'casuistique') might be worthy of consideration[13].

Similarly, contemporary medicine relies on the knowledge and mastery of the application of general principles with respect to individuals, communities and specific situations.

Kopelman states, however, and rightly so, that case studies and casuistry do not allow generalizations, and that they risk exposing the casuist researcher to biases favouring a particular individual or situation[14,15].

2.1.2 Casuistics in general understanding and daily life

The excessive detailing of cases, as undertaken by the Jesuits, was not always considered impartial and objective. This led to the image of a casuist as someone who argues a point in too subtle a manner. For many, casuistics or casuistry represented an excessively subtle process of argumentation[8,9].

Clinical casuistics or case reporting is quite a different story, as the rest of this text will illustrate.

2.1.3 General overview of single case concepts and design

Single case studies or single case research designs are often referred to in the general literature as intra-subject-replication designs, n-of-1 research, intensive designs, and so on[16].

This concept stems from a feeling that the advance of science should be based on the integration of what was acquired as information from one individual case (*idiographic approach*) as well as knowledge from a set of cases (*nomothetic approach*)[17].

The more general knowledge is applied to an individual, the more the appropriateness of the qualitative case methods should be considered[17].

A 'case' may be an individual, a given situation, an occurrence or an event in a particular area of daily or professional life. Case studies are carried out in many settings, such as in policy and political science, public administration, community psychology and sociology, organizational and management studies, and city and regional planning[18]. A process, a programme, a neighbourhood, an institution or its functioning, or a political, social, cultural or health event can be considered a case in the widest sense of the word. This term's exact meaning is still open to philosophical and conceptual discussion[19].

By stressing these concepts, it should be clear that our clinical case reports are part of a larger philosophical and scientific field than that of medicine and other health sciences.

As for clinical case reporting, the following recurring theme in science fiction movies and literature describes the situation quite well: we are not alone in the universe.

2.1.4 Case studies in administration, management, economics and business

The case method is used to solve a particular case rather than to provide an expanded answer to a more general question. In medicine, we want to work with the data in a patient's chart, to analyse it, and to give direction concerning the best clinical action. In business, the case is a situation to understand and make decisions about to improve a well-defined organization. Typical or extreme cases are analysed in detail as a way to learn how to solve problems[15].

Cases may be *empiric* (stemming from general experience) or *experimental* (the result of a deliberate manipulation).

2.1.5 Case studies in social sciences

Case studies owe their creation not only to medicine, but also to psychology, anthropology and other fields. Social sciences have perhaps made the greatest contributions to the general case study methodology, offering a wide world of learning[19–22] to the public at large.

2.1.5.1 Qualitative research

Contemporary medical research relies heavily on epidemiology and biostatistics. Observations are recorded, summarized, analysed and interpreted with care in order to avoid bias, random errors, or misrepresentation of the events or target groups being studied.

An impressive volume of concepts, methods and thinking has been discussed in the *ad hoc* literature[23-42]. Many experts in the social sciences stress the importance of establishing an equilibrium between what is found in *qualitative research,* i.e. an in-depth study of an individual or situation, and *quantitative research,* which summarizes sets of individual experiences.

Qualitative research was probably best defined by Strauss and Corbin[25] as '*any kind of research that produces findings not arrived at by means of statistical procedures or other means of quantification. It can refer to research about persons' lives, stories, behavior, but also about organizational, functional or social movements, or interactional relationships. Some data may be quantified as with census data but the analysis itself is a qualitative one'.*

In **quantitative research**, series of observations are made, and phenomena are quantified, counted, measured, described, displayed and analysed by statistical methods. Then, the results of these endeavours are applied to the problem as a whole. Biostatistics and epidemiology, which are representative of such undertakings in medicine, have been (and remain) pivotal in decision-oriented medical research. The analysis of disease outbreaks, clinical trials and systematic reviews (meta-analyses) of disease causes or treatment effectiveness are examples of quantitative research. In reality, the primary objective of quantitative research is *to provide answers to questions that extend beyond a single observation.*

In **qualitative research**, unique observations are the focus of interest. They are described, studied and analysed in depth. They are not considered as a representative part of a given field before being linked to other observations. The objective of this kind of research is primarily *to understand the case – a single observation itself.*

Research questions can be approached by induction or deduction. **Inductive research** proceeds from observations that serve as a basis for hypotheses and answers. **Deductive research** raises questions first, then gathers observations relevant to the problem, and confirms or rejects hypotheses 'free from or independent of the material under study'.

Quantitative research is both inductive and deductive. Epidemiologists, for obvious reasons, prefer the deductive component. Qualitative research is mostly inductive by definition. It is driven not by hypotheses, but by questions, issues and a search for patterns[43]. Classification is therefore its purpose.

We agree with Pope and Mays[44] that the focus of qualitative research is *'what is a given observation (X, case)?'*. Quantitative research, however, counts *'how many Xs?'*

Let us remember that qualitative research principles are not foreign to medicine. Medical histories, psychiatric interviews and the study of index cases (i.e. those that lead to quantitative research) of disease outbreaks or of new phenomena (poisoning, infection etc.) possess characteristics of qualitative research.

Obviously, equilibrium between quantitative and qualitative research is both necessary and beneficial[25,45–48], not only in social sciences but in health sciences as well.

2.1.5.2 Concept of a 'case'

The definition of the term 'case', which some[45] also refer to as 'site', remains slightly unclear[49]. It may[23,50] include the following (with examples given in parentheses):

- a person (in medicine and nursing);
- an event or situation (marital discord, strike);
- an action (spousal abuse) – and
- its remedy (consulting and its result);
- a programme (home care for chronic patients);
- a time period (baby boomers era);
- a critical incident (hostage taking);

- a small group (homeless people);
- a department (within an organization);
- an organization itself (company, political party);
- a community (people living on welfare, suburbanites etc.).

One 'case' may be studied and presented in different ways. For example, we can examine the Watergate case by focusing on the men behind the case, the event itself (a break-in at the office of a political party), or the attempts made to cover it up[18].

2.1.5.3 Case studies

> For Rothe[50] *'The term "case study" comes from the tradition of legal, medical and psychological research, where it refers to a detailed analysis of an individual case, and explains the dynamics and pathology of a given disease, crime or disorder. The assumption underlying case studies is that we can properly acquire knowledge of a phenomenon from intense exploration of a single example'.*

Case studies can be qualitative (based on a search for a meaning) or quantitative (based on some kind of measurement).

A clinical examination is both qualitative (history taking, psychiatric evaluation) and quantitative (anthropometry, examination of vital functions etc.).

In contrast, a social worker focuses more on qualitative information through three types of case studies[51]:

1. In an *intrinsic case study*, a better understanding of a case is sought (e.g. why did parents abandon their child?).

2. In an *instrumental case study*, a better understanding of the problem represented by the case or the refinement of an underlying theory of the problem is sought (e.g. parental child neglect).

3. A *collective case study* is not an epidemiological exploration (no target groups, no denominators etc.), but rather an extension of an instrumental case study of several cases. Its concept resembles that of case series reports, as we will see in Chapter 3.

Some researchers may study only selected phenomena that illustrate the case[52] adequately. Others may choose a *monographic study* that is as detailed and complete as possible. The nature of this kind of research is inductive[54].

Casuists in medicine and social sciences both agree that case studies do not permit generalizations. However, if a case study brings results contrary to previous experience, generalizations from previous experiences should be reviewed.

2.1.6 Other uses of single cases in biology and human sciences

In experimental psychology and physiology, studies of individual cases and sets of cases are undertaken to acquire new knowledge. In a quantitative study of multiple cases, the results obtained apply to groups or communities that are represented by the subjects being studied. They do not necessarily apply to each and every particular case (individual) under study.

To better understand what happens in a particular case, a *single case experimental design* is used[18,50,55,56]. This type of study should lead to improved decisions related to a particular individual or experimental subject studied. The *n-of-1 stu-*

dies in medicine (see Section 2.1.7) have a similar conceptual basis.

2.1.7 Return of qualitative research to medicine, nursing and public health

Qualitative research, like any other scientific concept, technique or method, isn't immune to the ebb and flow of information from one field to another.

In the past, experience gathered from the study of single individuals in medicine, psychology or nursing was used, modified and expanded in social sciences, business, finance, administration, law and the military. In these other areas, the 'case' concept extended beyond an individual to encompass situations, states or events. The resulting enriched methodology has now returned to the health sciences.

2.1.7.1 Qualitative approach in case research

Recently, the *BMJ* introduced its readers to qualitative research in health sciences, and in medicine in particular, by means of an excellent reader-friendly group of articles[44,57–63] that have appeared subsequently in book form[64].

As a result of impressive papers based on qualitative research, medical authors and readers now face increasing pressure with regards to the content and form of their written work. Greenhalgh[65] has suggested ways to read and understand *'papers that go beyond numbers'*. Hence, qualitative research has slowly joined the mainstream of findings in health sciences.

Family medicine calls for qualitative research[66]. Nursing[48,67–70], as a health profession and field of health research, has a longer history of qualitative research, for which it traditionally reserved a place of significant importance.

In health administration, medical care organization and health services research, case studies are used in their expanded sense[63]. Incentives arising from the removal from insurance of *in vitro* fertilization[71], the consequences of the introduction of public funding for midwifery[72], or patients' unmet expectations of care[73,74] can all be cited as examples.

2.1.7.2 Quantitative approach in case research

Patients as 'cases'

In the same spirit that encouraged experimental research in psychology, experimental research and clinical trials involving a single individual (*n-of-1 study, single case study*)[75–80] have been presented to physicians as a means of selecting the most appropriate treatment for a particular patient.

In this type of case research, instead of randomizing patients (there is only one anyway), treatment modalities are randomized. Multiple spells of disease of short duration with short remissions between them are particularly suitable for such an evaluation (angina, tension headache, migraine etc.). This approach can help physicians make decisions in individual cases. It can also be considered in cases where other types of clinical trials are impossible, for whatever reason, or in cases where the results of classical trials do not apply to the patient in question[75].

Situations as 'cases'

A 'case' may be a question to answer or, most commonly, a controversial[80] problem to clarify. For example, it might be of interest to better understand the effects of dietary intake or drug control on blood cholesterol levels. *Meta-analysis* as a kind of systematic review of evidence relating to the 'case' is generally very helpful in assessing these situations. In a wide sense, meta-analysis is a 'case study'.

In reality, the study of a clinical case (patient) should lead to the re-evaluation of acquired knowledge. It should offer something new, particularly in cases where other approaches cannot be used in an unbiased way and without a preconceived idea of the problem. The study of a clinical case must therefore be a 'systematic review' of evidence, even if the evidence is relatively weak at this level.

In the narrower area of an *'acquaintance with particulars'* [81], the study and reporting of clinical cases as medical casuistics remain the core practice of repeated daily use, and the most frequent method of qualitative research. The rest of this book is devoted to this subject.

2.2 MEDICAL CASUISTICS AS AN ENTRY LINK IN THE CHAIN OF EVIDENCE

Recent impressive developments in medical research based on increasingly large samples of individuals have made the study of single clinical cases a kind of neglected orphan of medical research rather than a necessity of daily clinical routine. Perhaps the opposite should occur.

Improved selection, observation, analysis, interpretation and reporting of a clinical case is a 'missing link' in the acquisition of medical knowledge.

Any advanced research should and often does begin with the study of one of several index cases. Figure 2.2 illustrates the situation.

Most often, we record the first case(s), i.e. the index case(s), of an interesting and important phenomenon. An **index case** is defined in medical genetics as an *'original patient (propositus or proband) which provides the stimulus for study of other members of the family, to ascertain a possible genetic factor in causation of the presenting condition'* [4]. By extension, any first case that triggers a more advanced investigation is an index case.

First cases of vomiting and diarrhoea might be index cases of a food poisoning outbreak, leading to the investigation of an epidemic. In toxicology and environmental medicine, the first occurrences of respiratory problems led to an elucidation of the role of toxic fumes in silo filler's disease. In the area of new medical technology, the first unexpected deaths of some

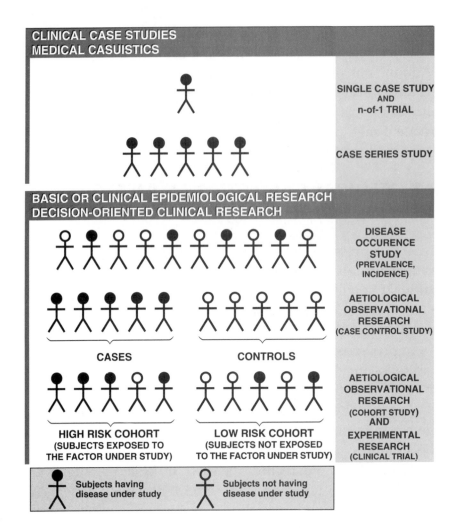

Figure 2.2 – Medical research and lines of evidence according to individuals involved.

users of an insulin pump were index cases leading to the surveillance of the problem, the evaluation of the risk factors, and ultimately to the modification of the criteria for initiating pump use[82]. Any unexpected beneficial effect of a drug in situations other than those for which it was primarily intended forms an index case for an *ad hoc* clinical trial. As an example, Minoxidil was found to be a hair growth stimulant and the intended use of Viagra®, changed from the treatment of hypertension.

Figure 2.2 outlines the cascading steps of a medical investigation:

Step 1. Case studies are the starting point in the development of a portrait of disease occurrence, its treatment or outcomes, based on epidemiological surveillance and other means.

Step 2. Linking cases to one or more control groups (unaffected individuals) means that a case control search for causes will be performed. A group of unaffected individuals is added to a set of cases.

Step 3. Comparing exposed and unexposed individuals produces a cohort study, be it observational or experimental (clinical trial). Cases will develop or improve depending on exposure (drug, poison etc.).

If a case study means 'noticing the problem', such noticing must take place at other steps of the cascade as well. Today, the methodology of case studies and reports must be as refined and sophisticated as that of case control studies or clinical trials.

In addition, because a case study is a first link in the chain of evidence, other steps do not necessarily have to follow for some time. A single case or case series (with all their inherent limitations) may long remain the only evidence available. If that happens, single cases or case series must provide the best evidence in their contexts. Long ago, generations of faithful were naturally curious about the Old and New Testament's 'case reports' of various miracles and heavenly events. To this day, like the Immaculate Conception and Ezekiel's bones, they remain the only cases universally known.

If we are reasonably well trained and experienced in aetiological observational research and clinical trials, should we not master equally well the methodology of clinical case study and reporting?

2.3 IS THERE A NEED FOR BETTER TRAINING IN CLINICAL CASE STUDIES – AND HOW IMPORTANT ARE CLINICAL CASE REPORTS TODAY?

2.3.1 Medical casuistics

Case studies and qualitative research share many characteristics. As soon as a patient enters a medical office or hospital, the attending physician, often without knowing it, uses one or several components of qualitative research. Clinimetrics are a quantitative counterpart in the study and assessment of clinical cases.

Moreover, qualitative aspects of care and research differentiate such disciplines as family medicine or psychiatry. Psychoanalysis *per se* has been a kind of qualitative research since Josef Breuer and Sigmund Freud's 'Case of Anna O (Bertha Pappenheim)'.

We are all aware that a better understanding of many previously poorly known phenomena was made possible by the relevant observation and reporting of index cases. Examples are the case of maple syrup disease, isoimmunization in pregnancy, acquired immunodeficiency syndrome and multiple personality disorder, to name just a few. Some clinical case reports pertaining to these illnesses have become landmark articles in medicine.

William Osler knew that the best training in medicine, as a whole, is that which focuses on individual cases and bedside care[83]: *'The amphitheatre clinic, the ward and dispensary classes are but bastard substitutes for a system which makes the medical student himself help in the work . . . put him (i.e. the student) behind a case. Ask any physician of twenty years' standing how he has become proficient in his art, and he will reply that (what) he learned in the schools was totally different from the medicine he learned at the bedside. Medicine is learned by the bedside and not in the classroom'.* If such experiences are not properly recorded and shared through a clinical case report, for instance, the most valuable part of clinical wisdom will be lost.

As Kathryn Montgomery Hunter stated[21]: *'Despite the refinements of clinical epidemiology and the study of decision making, the careful construction and interpretation of the individual case remains essential to learning and remembering in clinical medicine. Medical education cultivates and clinical practice refines clinical judgement. Medicine, then, is practised casuistically somewhere between the forest of textbooks and a thicket of (algorithm) trees. Narrative is the ultimate device of casuistry in medicine (as in theology and law), which enables practitioners who share its diagnostic and therapeutic world view to fit general principles to the single case and to achieve a degree of generalization that is both practicable and open to change'.*

If clinical case study and reporting regroups the qualitative finesse of observation, the rigor of measurement and classification of clinimetrics, and the pragmatic need for ensuing medical decisions, then medical casuistics is not only *'the recording and study of cases of any disease'* [1], but also **the art of choosing, gathering, structuring and conveying pragmatic information about relevant clinical cases. It must lead to a better understanding of a given health problem and improved clinical decisions. It should focus on the reduction of information to be gathered, rather than on the accumulation and detailing of a given volume of data.**

Clinical medicine needs direction with regards to what to do with particular patients or groups of patients. Hence, *'The more a programme aims at individual outcomes, the greater the appropriateness of qualitative case methods. The more a programme emphasizes common outcomes for all participants, the greater may be the appropriateness of standardized quantitative measures of performance and change'* [17].

2.3.2 Specific situations and questions as cases

For some, a case study should evaluate only controversial situations. A 'case' then becomes a problem to be solved. As an example, the role of dietary intake or drugs on blood cholesterol levels can be studied. Meta-analysis can be used to clarify the problem.

In the same spirit, a unique case study should incite the revision of acquired knowledge. It should offer something new, for example an answer to a question that cannot be obtained by other types of studies or means. Therefore, shouldn't a systematic review of literature or a meta-analysis be seen as a kind of 'case study'?

Such a widened concept of medical casuistics is based on new paradigms and on the revision of old ones that focused on one particular case or on several cases taken from hospital or community settings.

2.4 CONCLUSIONS

Multiple cases call for a balance between qualitative and quantitative research in medicine and other health sciences.

A 'case' may be an individual patient, for example, whose unexpectedly localized aneurysm rupture is associated with some unexpected clinical manifestations possibly requiring some special emergency and surgical care.

A situation may be a 'case'. Recently, an in-flight emergency was reported[84,85]. A passenger developed a tension pneumothorax requiring an improvised chest drain on board an aircraft. The problem was solved by using improvised material including a modified coat hanger as a trocar, a bottle of mineral water with two holes created in its cap as an underwater seal drain, and five star brandy as a disinfectant for the introducer. The treatment was successful and a proposal for dealing with future similar air transportation emergencies was published in the medical press.

Case reports should not simply be centred on presenting unusual cases. They should lead to the surveillance of rare cases, to the explanation of the underlying mechanisms and treatment of cases, and to the direction required for future research[86]. They should also generate new hypotheses although these hypotheses cannot be confirmed by single cases.

A specific methodology must be taught and put into practice as outlined in the next chapter.

Is this really necessary? Absolutely!

Calls for qualitative research abound in various specialties[81,87–90], as do calls for better clinical case presentations[91–94] because they play an ever-increasing role in various clinical training programmes and are subject to evaluation[95–102]. Moreover, editors of medical journals now provide more detailed indications of their expectations with regards to clinical case reporters[103,104] and their reports.

REFERENCES

1. *Mosby's Medical Dictionary*, 5th Edition. St Louis: Mosby, 1998.
2. Jenicek M. *Casuistique Médicale. Bien Présenter un Cas Clinique.* (Medical Casuistics. Good Clinical Case Reporting.) St Hyacinthe and Paris: Edisem and Maloine, 1997.
3. Vereecke L-G. Casuistique. pp. 61–62 in: *Encyclopaedia Universalis*, Corpus 5. Paris: Encyclopaedia Universalis, Éditeur à Paris, 1989.
4. Dorland WAN. *Dorland's Illustrated Medical Dictionary*, 27th Edition. Edited by EJ Taylor. Philadelphia: WB Saunders, 1988.
5. *Collins Robert French/English English/French Dictionary (Unabridged)*, 3rd Edition. Glasgow: Harper Collins, 1993.

6. *Dictionnaire de Médecine Flammarion*, 4th Edition. Paris: Flammarion, 1991.

7. *Encyclopedia Britannica*, 15th Edition. Chicago: Encyclopedia Britannica, 1992.

8. *Dictionnaire de Français 'Plus' à l'Usage des Francophones d'Amérique*. Montreal: CEC (Centre Educatif et Culturel Inc.),1988.

9. *Dictionnaire Encyclopédique Larousse*. Paris: Librairie Larousse, 1979.

10. Jonsen AR. Casuistry and clinical ethics. *Theor Med*,1986;**7**:65–74.

11. Jonsen AR. Casuistry as methodology in clinical ethics. *Theor Med*,1991;**12**:295–307.

12. Arras JD. Gettting down to cases: The revival of casuistry in bioethics. *J Med Phil*,1991;**16**:29–51.

13. Jonsen AR, Toulmin S. *The Abuse of Casuistry. A History of Moral Reasoning*. Berkeley: University of California Press, 1988.

14. Kopelman LM. Case method and casuistry: The problem of bias. *Theor Med*,1994;**15**:21–37.

15. Pagès R. Cas (méthode des), pp. 34–37 in *Encyclopaedia Universalis*, Corpus 5. Paris, Encyclopaedia Universalis, Éditeur à Paris, 1989.

16. Kazdin AE. *Single-Case Research Designs. Methods for Clinical and Applied Settings*. New York: Oxford University press, 1980.

17. Patton MQ. *Qualitative Evaluation Methods*. Beverly Hills: Sage Publications, 1980.

18. Yin RK. *Case Study Research. Design and Methods*. Newbury Park: Sage Publications, 1988.

19. *What Is a Case? Exploring the Foundations of Social Inquiry*. Edited by CC Ragin and HS Becker. Cambridge: Cambridge University Press, 1992.

20. Barlow DH, Hersen M. *Single Case Experimental Designs. Strategies for Studying Behavior Change*. New York: Pergamon Press, 1984.

21. Hunter Montgomery K. A science of individuals: Medicine and casuistry. *J Med Phil*,1989;**14**:193–212.

22. Stake R. *The Art of Case Study Research*. Thousand Oaks: Sage, 1995.

23. Miles M, Huberman AM. *Qualitative Data Analysis: A Sourcebook of New Methods*. Newbury Park: Sage, 1989.

24. Guba EG, Lincoln YS. *Fourth Generation Evaluation*. Newbury Park: Sage, 1989.

25. Strauss A, Corbin J. *Basics of Qualitative Research. Grounded Theory Procedures and Techniques*. Newbury Park: Sage Publications, 1990.

26. Tesch R. *Qualitative Research: Analysis Types and Software Tools*. New York: The Falmer Press, 1990.

27. Patton MQ. *Qualitative Evaluation and Research Methods*. Newbury Park:Sage, 1990.

28. Crabtree BF, Miller WL. *Doing Qualitative Research*. Newbury Park: Sage, 1992.

29. Wolcott HF. *Transforming Qualitative Data*. Thousand Oaks: Sage, 1994.

30. Guba EG, Lincoln YS. Competing paradigms in qualitative research, pp. 105–119 in: *Handbook of Qualitative Research*. Edited by NK Denzin and YS Lincoln. Thousand Oaks: Sage, 1994.

31. Creswell JW. *Research design. Qualitative & Quantitative Approaches*. Thousand Oaks: Sage, 1994.

32. Miller WL, Crabtree BF. Clinical research, pp. 340–352 in: *Handbook of Qualitative Research*. Edited by NK Denzin and YS Lincoln. Thousand Oaks: Sage, 1994.

33. Adler PA, Adler P. Observational techniques, pp. 377–392 in: *Handbook of Qualitative Research*. Edited by NK Denzin and YS Lincoln. Thousand Oaks: Sage, 1994.

34. Fontana A, Frey JH. Interviewing. The art of science, pp. 361–376 in: *Handbook of Qualitative Research*. Edited by NK Denzin and YS Lincoln. Thousand Oaks: Sage, 1994.

35. Marshall C, Rossman CB. *Designing Qualitative Research*, 2nd Edition. Newbury Park:Sage, 1994.

36. Rubin HJ, Rubin IS. *Qualitative Interviewing. The Art of Hearing Data*. Thousand Oaks: Sage, 1995.

37. Weitzman EA, Miles MB. *A Software Sourcebook: Computer Programs for Qualitative Data Analysis*. Thousand Oaks: Sage, 1995.

38. Mason J. *Qualitative Researching*. London: Sage, 1996.

39. Kvale S. *Interviews: An Introduction to Qualitative Research Interviewing*. Thousand Oaks: Sage, 1996.

40. Coffey A, Atkinson P. *Making Sense of Qualitative Data*. Thousand Oaks: Sage, 1997.

41. Schwandt TS. *Qualitative Inquiry: A Dictionary of Terms*. Thousand Oaks: Sage, 1997.

42. Creswell JW. *Qualitative Inquiry and Research Design. Choosing Among Five Traditions*. Thousand Oaks: Sage, 1998.

43. Patton MQ. *How to Use Qualitative Methods in Evaluation*. Newbury Park: Sage Publications, 1987.

44. Pope C, Mays N. Reaching the parts other cannot reach: an introduction to qualitative methods in health and health services research. *BMJ*,1995;**311**:42–5.

45. Denzin NK, Lincoln YS. Introduction. Entering the field of qualitative research (Chapter 1), pp. 1–17 in: *Handbook of Qualitative Research*. Edited by NK Denzin and YS Lincoln. Thousand Oaks: Sage, 1994.

46. Miles MB, Huberman AM. *Qualitative Data Analysis: A Sourcebook of New Methods*. Beverly Hills: Sage Publications, 1984.

47. Verhoef MJ, Casebeer AL. Broadening horizons: Integrating quantitative and qualitatve research. *Can J Infect Dis*,1997;**8**:65–6.

48. Dzurec LC, Abraham IL. The nature of inquiry: Linking quantitative and qualitative research. *Adv Nurs Sci*,1993;**16**:73–9.

49. Ragin CC. Introduction: Cases of 'What is a case?', pp. 1–17 in: *What is a Case? Exploring the Foundation of Social Inquiry*. Edited by CC Ragin and HS Becker. Cambridge: Cambridge University Press, 1992.

50. Rothe JP. *Qualitative Research. A Practical Guide*. Heidelberg and Toronto: RCI/PDE Publications, 1993.

51. Stake RE. Case studies (Chapter 14), pp. 236–247 in: *Handbook of Qualitative Research*. Edited by NK Denzin and YS Lincoln. Thousand Oaks: Sage Publications, 1994.

52. Aday LA. *Designing and Conducting Health Surveys. A Comprehensive Guide*. San Francisco and London: Jossey-Bass Publishers, 1989.

53. Patton MQ. *How to Use Qualitative Methods in Evaluation*. Newbury Park: Sage Publications, 1987.

54. Yin RK. *Applications of Case Studies Research*. Thousand Oaks: Sage Publications, 1993.

55. Kazdin AE. *Single-Case Research Designs. Methods for Clinical and Applied Settings*. New York: Oxford University Press, 1982.

56. Barlow DH, Hersen M. *Single Case Experimental Designs. Strategies for Studying Behavior Change*. New York: Pergamon Press,1984.

57. Jones R. Why do qualitative research? It should begin to close the gap between the sciences of discovery and implementation. *BMJ*,1995;**311**:2.

58. Mays N, Pope C. Rigour and qualitative research. *BMJ*,1995;**311**:109–12.

59. Mays N, Pope C. Observational methods in health care settings. *BMJ*, 1995;**311**:182–4.

60. Britten N. Qualitative interview in medical research. *BMJ*,1995;**311**:251–3.

61. Kitzinger J. Introducing focus groups. *BMJ*,1995;**311**:299–302.

62. Jones J, Hunter D. Consensus methods for medical and health services research. *BMJ*,1995;**311**:376–80.

63. Keen J, Packwood T. Case study evaluation. *BMJ*,1995;**311**:444–6.

64. *Qualitative Research in Health Care*. Edited by N Mays and C Pope. London: BMJ Publishing Group, 1996.

65. Greenhalgh T. Papers that go beyond numbers (qualitative research) (Chapter 11), pp. 151–162 in: *How to Read a Paper. The Basics of Evidence Based Medicine*. London: BMJ Publishing Group, 1997.

66. Burkett GL, Godkin MA. Qualitative research in family medicine. *J Family Practice*,1983;**16**:625–6.

67. Goodwin LD, Goodwin WL. Qualitative *vs.* quantitative research or qualitative *and* quantitative research? *Nurs Res*,1984;**33**:378–80.

68. Skodol Wilson H. Qualitative studies: from observations to explanations. *J Nurs Admin*,1985;**15**:8–10 (May).

69. Meier P, Pugh EJ. The case study: A viable approach to clinical research. *Res Nurs Health*,1986;**9**:195–202.

70. Sandelowski M. One of the liveliest number: The case orientation of qualitative research. *Res Nurs Health*,1996;**19**:525–9.

71. Giacomini M, Hurley J, Stoddart G, Schneider D, West S. When tinkering is too much. A case study of incentives arising from Ontario's deinsurance of *in vitro* fertilization. Paper 96-10. Hamilton: McMaster University Centre for Health Economics and Policy Analysis, Working Paper Series, August 1996.

72. Giacomini M, Peters M. The introduction of public funding for midwifery in Ontario: Interpreting the meaning of the financial incentives. Paper 96-9.

Hamiton: McMaster University Centre for Health Economics and Policy Analysis, Working Paper Series, August 1996.

73. Kravitz RL, Callahan EJ, Paterniti D, Antonius D, Dunham M, Lewis CE, Prevalence and sources of patients' unmet expectations for care. *Ann Intern Med*,1996;**125**:730–7.

74. Inui TS. The virtue of qualitative *and* quantitative research. *Ann Intern Med*, 1996; **125**:770–1.

75. Gyuatt G, Sackett D, Taylor DW, Chong J, Roberts R, Pugsley S. Determining optimal therapy – randomized trials in individual patients. *N Engl J Med*, 1986;**314**:889–92.

76. Guyatt G, Sackett D, Adachi J, Roberts R, Chong J, Rosenbloom D, Keller J. A clincian's guide for conducting randomized trials in individual patients. *CMAJ*,1988;**139**:497–503.

77. Single Case Studies. Proceedings from a symposium in Oslo, 14–15 March 1987. Edited by H Petersen. *Scand J Gastroenterol*,1988;**23**:Suppl 147,1–48.

78. Guyatt GH, Keller JL, Jaeschke R, Rosenblooom D, Adachi JD, Newhouse MT. The n-of-1 randomized controlled clinical trial: Clinical usefulness. Our three year experience. *Ann Intern Med*,1990;**112**:293–9.

79. Hodgson M. N-of-one clinical trials. The practice of environmental and occupational medicine. *J Occup Environ Med*,1993;**35**:375–80.

80. West R. Assessment of evidence versus consensus or prejudice. *J Epidemiol Comm Med*, 1992;**46**:321–2.

81. McWhinney IR. 'An acquaintance with particulars'. *Fam Med*,1989;**21**:296–8.

82. Thacker SB, Berkelman RL. Surveillance of medical technologies. *J Public Health Policy*,1986;**7**:363–77.

83. *Counsels and Ideals from the Writings of William Osler & Selected Aphorisms.* Compiled and edited by CNB Camac (Counsels and Ideals) and LJ Beam (Selected Aphorisms). Birmingham: The Classics of Medicine Library, 1985.

84. Wallace WA. Managing in flight emergencies. A personal account. *BMJ*, 1995;**311**:374–6.

85. Wallace WA, Wong T, O'Bichere A, Ellis BW. Discussion, *BMJ*, 1995;**311**: 375–6.

86. Fletcher RH, Fletcher SW, Wagner EH. Studying cases (Chapter 10), pp. 208–212 in: *Clinical Epidemiology. The Essentials*. Baltimore: Williams & Wilkins, 1996.

87. McWhinney IR. Changing models: The impact of Kuhn's theory on medicine. *Family Practice*,1983;**1**:3–8.

88. Rainsberry RPN. Values, paradigms and research in family medicine. *Family Practice*,1986;**3**:209–15.

89. Naess MH, Malterud K. Patients' stories: science, clinical facts or fairy tales? *Scand J Prim Health Care*,1994;**12**:59–64.

90. Shafer A, Fish MP. A call for narrative: The patient's story and anesthesia training. *Literature and Medicine*,1994 (Spring);**13**:124–42.

91. Engel GL. The deficiencies of the case presentation as a method of clinical teaching. *N Engl J Med*,1971;**284**:20–4.

92. Allen LC, Bunting PS. Postdoctoral training in clinical chemistry: Laboratory training aspects. *Clin Biochem,*1995;**28**:481–97.

93. Hormbrey P, Todd BS, Mansfield CD, Skinner DV. A survey of teaching and the use of clinical guidelines in accident and emergency departments. *J Accid Emerg Med,*1996;**13**:129–33.

94. Haddad D, Robertson KJ, Cockburn F, Helms P, McIntosh N, Olver RE. What is core? Guidelines for the core curriculum in paediatrics. *Med Educ,*1997;**31**:354–8.

95. Klos M, Reuler JB, Nardone DA, Girard DE. An evaluation of trainee performance in the case presentation. *J Med Educ,*1983;**58**:432–4.

96. Kihm JT, Brown JT, Divine GW, Linzer M. Quantitative analysis of the outpatient oral case presentation: Piloting a method. *J Gen Intern Med,*1991;**6**:233–6.

97. Gennis VM, Gennis MA. Supervision in the outpatient clinic: Effects on teaching and patient care. *J Gen Intern Med,*1993;**8**:378–80.

98. Templeton B, Selarnick HS. Evaluating consultation psychiatry residents. *Gen Hosp Psychiatry,*1994;**16**:326–34.

99. Raik B, Fein O, Wachspress S. Measuring the use of the population perspective on internal medicine attending rounds. *Acad Med,*1995;**70**:1047–9.

100. Greenberg LW, Getson PR. Assessing student performance on a pediatric clerkship. *Arch Pediatr Adolesc Med,*1996;**150**:1209–12.

101. Elliot DL, Hickam DH. How do faculty evaluate student's case presentations? *Teach Learn Med,*1997;**9**:261–3.

102. Bass EB, Fortin AH, Morrison G, Wills S, Mumford LM, Goroll AH. National survey of clerkship directors in internal medicine on the competencies that should be addressed in the medicine core clerkship. *Am J Med,*1997;**102**:564–71.

103. Squires BP. Case reports: What editors want from authors and peer reviewers. *CMAJ,*1989;**141**:379–80.

104. Squires BP, Elmslie TJ. Reports of case series: What editors expect from authors and peer reviewers. *CMAJ,*1990;**142**:1205–6.

CHAPTER 3

How to prepare a single case report

How to prepare a single case report

How many times have we heard our senior colleagues declare: 'since he can't do clinical research, let him report clinical cases'? Yet the case study is an important and integral part of clinical research. It is our duty to grant it its own letters of nobility, by providing it with a logic, a methodology, an architecture and a series of objectives comparable in their rigour to those of other types of research in medicine: aetiological research, evaluation of diagnostic methods, clinical trials, and studies of prognosis.

A clinical case report is a form of verbal or written communication with its own specific rules, that is produced for professional and scientific purposes. It usually focuses on an unusual single event

> (patient or clinical situation) in order to provide a
> better understanding of the case and of its effects
> on improved clinical decision-making.

In their instructions to contributors, medical journals generally specify the required length of text (number of words) and newness of observation and nothing more. However, there are many other guidelines to follow.

The study and reporting of cases in clinical practice occurs at two different levels, as outlined below.

1. **Routine case reports** (admissions and discharges) ensure the continuity of ward activities or outline those activities directly related to the proper functioning of the hospital.
2. **Case reports of scientific value**, usually topics of grand rounds and reports in medical journals, focus on:
   - case reports;
   - case reports with a literature review;
   - case series reports;
   - systematic reviews of cases.

This second category presents the most challenges.

There are common rules for all the above-mentioned categories of clinical case reports. Both general explanations of these rules and specific recommendations will follow in this text, which focuses mainly on clinical case reports of scientific value.

Do we really need to investigate this area? Should we not be satisfied with the existing literature?

In reality, the bibliographic heritage of medical casuistics is not very rich, especially that pertaining to case reporting methodology. Aside from clinical epidemiological investigations of case reports, the past twenty years have generally only produced 'how to' articles covering either routine ward case reporting[1-4], the content[5-8] and form[9-12] of the case report, or strategies to adopt in order to ensure the case report's publication[13,14].

More recently, however, editors have begun to state their expectations more precisely[5,15-17], although some of these expectations differ from the supporting literature and some aren't backed up by the available methodology in the medical press. In this chapter, we will attempt to create a stronger bond between the content and the form of the product in order to meet consumer expectations.

As already mentioned, clinical case reports must be seen as the first link in the chain of evidence. Reliance on clinical epidemiology, clinimetrics and an evidence-based approach to medical information requires additional considerations. How, without epidemiology, can a case reporter qualify the risk and prognostic characteristics of his case? How, without clinimetrics, can the clinical and paraclinical features of a case be presented in operational and measurable terms? How can the evidence (weak or not) that emerges from a case be compared with other evidence from fields related to the case?

The case report must be as solid as a rock or as the other pieces of evidence associated with the problem under study.

3.1 ROUTINE WARD CASE REPORTS AND HOSPITAL-FOCUSED CASE PRESENTATIONS

Routine case reports are the most ubiquitous of all single case reports. Rather than a verbatim retelling of the admission work-up, they should represent a 'medical reporting'[1], a succinct account of an event.

Such case presentations usually follow an expanded 'S.O.A.P.' structure[3], which should be more than familiar to any North American house staff member preparing daily progress notes on his patients:

- Presenting the problem.
- Subjective history of present illness, and personal and family history.
- Objective data (clinical and paraclinical, course so far).
- Assessment (diagnosis, differential diagnosis, co-morbidity review).
- Plans of treatment and care, including their evaluation in terms of material and human resources (management of the case).
- Discussion.

In psychiatry, important additions are made:
- Objective observation and assessment of behaviour.
- Evaluation of mood, affect and thought.
- Sensory perception and functioning.
- Subjective and objective insight.

We should note that a case work-up and report are not only the result of a properly conducted interview[4,18] (including review of systems), but also a structured synthesis of additional elements such as paraclinical work-ups and results, outcome assessments and social evaluations (e.g. reintegration into the patient's family, professional and social environment).

A much greater challenge is presented when cases are reported for the advancement of knowledge and science.

3.2 CLINICAL CASE REPORTING AS A SCIENTIFIC CONTRIBUTION TO EVIDENCE IN MEDICINE

Let us stress again that a clinical case report is a form of scientific and professional communication. As such, it has its own rules. Surprisingly, most medical journals give prospective authors no indications other than those related to the newness of the topic and the volume of the text (number of words). More details are clearly needed.

The success of a clinical case report depends on four major criteria:
- the relevance of the topic;
- the value of the presentation (precision, organization, structure);
- the knowledge that this case contributes to the existing general information about the topic;
- the clarity of the elements that can be retained for practice, research, or both.

Hence, a good clinical case report must:
- be written according to the preset requirements;
- contain all essential elements, in a complete and well-structured manner;
- convey an unambiguous message.

Readers who believe that they do not need clinical epidemiology and biostatistics to present a single case, may have to think again. Even a particular case should be analysed in the context of the pathology it represents and in the framework of a particular clinical practice. In one way or another, the case should be compared with what is considered usual and general.

3.2.1 Single case reports

Single case reports fall into several categories:
- Either a *'classical case report'* is produced, where all necessary components and the discussion are limited to the case and the problem it represents, or
- a case is presented as a *'brief report'* in journals such as *The Lancet*, with only the most essential elements published (four typed pages or a half-page printed), or

- *'a case to learn from'* is prepared. The *New England Journal of Medicine* has created two sections for this type of case. In the Clinical Problem Solving section, the case is presented step by step by a seasoned clinician. The author shares with the reader the experience of going from one stage of the report to another. In the Case Records of the *Massachusetts General Hospital Weekly* Clinicopathological Exercises, the case is presented in its entirety. A pathologist then re-evaluates it.

3.2.2 Requirements and expectations regarding clinical case reports

Any successful clinical case reporter must keep in mind several considerations, desirable attributes and necessary components of a good report before sitting down to prepare it. The following suggestions are a reflection of the requirements and expectations laid out by medical journals for a good case report[15–17]. They should increase the author's chance of being published. The saying that your sleep will be only as good as the way you make your bed is valid here as well.

All case reports must be prepared with a specific reason in mind and must be based on high quality clinical data.

3.2.2.1 Reasons and motives for a clinical case report

When the time of year for promotions arrives, many clinicians feel a sudden urge to report clinical cases in order to enhance their list of publications. However, there are eighteen more serious reasons to publish a case report (see Table 3.1). These reasons should be stated in the introduction of all clinical case reports and, ideally, any one of these reasons should help to refocus clinical decisions.

However, everything depends on the relevance of the case. Riesenberg[18] notes that, from a recent selection of fifty-one landmark articles in medicine[19], five are *'just a simple'* case report. For example, Levine and Stetson published 'An unusual case of intra-group agglutination'[20]. This careful

Table 3.1 Reasons and motives for a case report

1. Unusual presentation of unknown aetiology.
2. Unusual natural history.
3. Unusual natural or clinical courses (spectrum, gradient, prognosis).
4. Challenging differential diagnosis.
5. Mistake in diagnosis, its causes and consequences.
6. Unusual and/or unexpected effect of treatment.
7. Diagnostic and therapeutic 'accidents' (causes, consequences, remedies).
8. Unusual co-morbidity (its diagnosis, treatment, outcome).
9. Transfer of medical technology (disease, organ, system).
10. Unusual setting of medical care.
11. Management of an emergency case.
12. Patient compliance.
13. Patient/doctor interaction (as in psychiatry).
14. Single case clinical trial ('n-of-one' study).
15. Clinical situation which cannot be reproduced for ethical reasons.
16. Limited access to cases.
17. New medical technology (use, outcomes, consequences).
18. Confirmation of something already known (only if useful for a 'systematic case report review and synthesis').

observation of one single case led the authors to conclude[21] that the majority of cases of erythroblastosis fetalis resulted from isoimmunization of an Rh-negative mother by the Rh-positive red blood cells of the fetus. Greenwalt mentions that *some may scoff at the publication of case reports, but for the astute scientist, a carefully documented study of an unusual patient represents an experiment of nature that may be the opportunity to explain a long-recorded but unexplained clinical mystery. Some credit must also go to the editorial staff of The Journal for having published this'* [22].

Sometimes, the message of the case report is missed because the reader does not know the desirable attributes of a good clinical case report, how to interpret a case report, and where extrapolations, generalizations and applications of a report should stop. All efforts made by the authors and the publishing journal would be futile in this situation.

3.2.2.2 Selection and quality of the clinical and paraclinical data on which a report is based

Selection of data

It is impossible to reproduce everything that a clinician has seen. Editorial space is limited.

The objective of the case report is not to prove that the clinician has done his job properly but rather to offer all information necessary for the understanding of the problem illustrated by the case. A 'classical' casuist searches for and provides all in-depth details of the case. A clinical casuist 'goes for the jugular' by providing just the essential information[23].

Data should be provided to allow the reader to understand the other steps of the case reporter's work, the differential diagnosis, or the choice of the treatment. A case report on a complicated excision of a cyst in an unusual and surgically challenging part of the inguinal region does not necessarily require an explanation of the patient's normal chest X-ray or glycaemia.

Quality of data

Quality of data counts even if it is not evident in a brief report itself.

A case report provides *clinical data* (for example, that a patient's blood pressure is 180/110 mmHg) as well as *clinical information* (the fact that the patient suffers from hypertension) as an interpretation of raw observations (data).

The reporter must have at hand and be ready to explain, on request, his measurement techniques and his inclusion and exclusion criteria for given information, readings or recording in *operational* terms. *Conceptual* criteria are not enough.

Soft data such as nausea, pain, anxiety, loneliness, puffiness and swelling, so well known to family doctors or psychiatrists, represent a special challenge. Their definitions should be as close as possible to those of **hard data**. Hard data come largely from the paraclinical (laboratory) area: blood count, urinary output, ventricular ejection fraction etc. Methods for *hardening of soft data* are discussed in the current epidemiological literature[24–26]. For example, measurement of the severity of pain may be attempted using an appropriate scale, or a degree of confusion in an elderly patient may be evaluated by an *ad hoc* psychiatric questionnaire.

A clinical case reporter should keep a record of his data collection and interpretation, in order to offer adequate explanations, should the need arise.

A well-presented case based on valid data follows the same rules as any other research topic would[27].

3.2.2.3 Content and structure of a single case report

A journal's space and requirements permitting, a clinical case report should have five distinct sections (see Table 3.2).

3.2.2.3.1 Title

Two kinds of titles are found in the literature. The first are symbolic or poetic titles, some of which possess advertising qualities. For example, a title like 'The rooster that sang a different song' as an introduction to a case report on an unusual clinical picture of whooping cough may perhaps reflect the author's wit and draw attention. However, the reader might have to read through the whole report in order to

Table 3.2

The five sections of a clinical case report

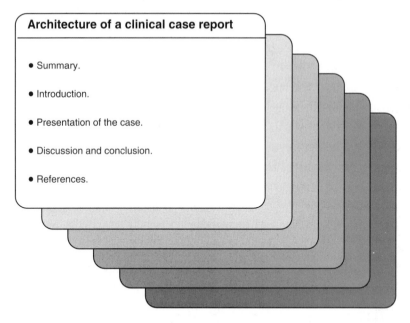

Architecture of a clinical case report

- Summary.
- Introduction.
- Presentation of the case.
- Discussion and conclusion.
- References.

understand what it is about. If, as in Agatha Christie's or Sir Arthur Conan Doyle's novels, the mystery is unravelled only at the end of the clinical case report, its reading is often painful. Usually this type of structure suggests that the author wanted to say: 'look how clever I am'.

The other kind of title is one that directly informs the readers about the problem and the topic. It conveys the elements of a well-formulated research question. In original research or in systematic reviews, research questions should reflect the following train of thoughts:

intervention → outcomes → population setting and condition of interest[28].

For example[28]:

'Does **anticoagulation therapy** improve **outcomes** in **patients** with **ischaemic stroke**?'

or[29]:

'Do **anticoagulant agents** improve **outcomes** in **patients with acute ischaemic stroke** compared with **no treatment**?'

Similar principles apply to titles of case reports. The informative title (or research question) is definitely preferable for a clinical case report. For example, the case report that is analysed and annotated in the next chapter is entitled **'Electrocardiographic changes** suggestive of **cardiac ischaemia** in a **patient** with **oesophageal food impaction'** [30]. The reader immediately knows the diagnostic intervention of interest (ECG), its outcome (changes suggestive of cardiac ischaemia), the population and the condition of interest (patient with oesophageal food impaction). He may decide if this topic interests him and whether he wants to read the rest. Only the subtitle is symbolic and allegorical, since it acts as an attention grabber: 'A case that's hard to swallow'. It should be noted that if the subtitle had been the main title, an uninformed reader might not have known what to expect. In other words, the title should always get right to the point.

3.2.2.3.2 Summary

A summary of any professional communication has two objectives:

- to attract the reader to the topic in a striking and organized manner; and
- to convey the most important highlights to a busy reader.

Many high impact journals including *The Lancet, JAMA, The New England Journal of Medicine* and *Annals of Internal Medicine* require well-structured summaries[31–33] for medical research articles. Independent of the nature of the article, a summary should give some background information about the problem, while also stating the objective, the design, the setting, the subjects, the results of the study and their meaning. Table 3.3 shows that a similar structure applies to the summary of a clinical case report.

Table 3.3
Organization of the summary

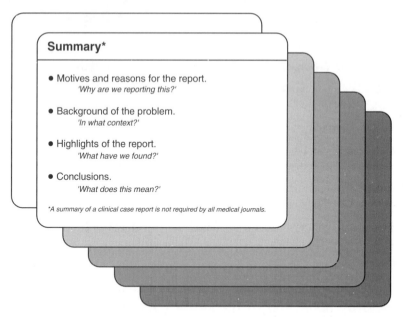

Summary*

- Motives and reasons for the report.
  'Why are we reporting this?'

- Background of the problem.
  'In what context?'

- Highlights of the report.
  'What have we found?'

- Conclusions.
  'What does this mean?'

*A summary of a clinical case report is not required by all medical journals.

For example, summaries constructed as above are expected to introduce clinical case reports in special Case Reports issues of the *Obstetrics and Gynecology* and *Revue de Pédiatrie* journals.

3.2.2.3.3 Introduction

In this section, the case has to be 'sold' to the reader, especially if there is no summary. It should persuade him or her to read through the whole text. It should also give all the necessary information about the problem under study.

Four kinds of information should be included in the introduction. Table 3.4 summarizes them.

Let us now examine in greater detail some components of the introduction.

Problem under study. The topic of a case report can be new[5,23,34,35] for the reasons already summarized in Table

Table 3.4

Organization of the introduction

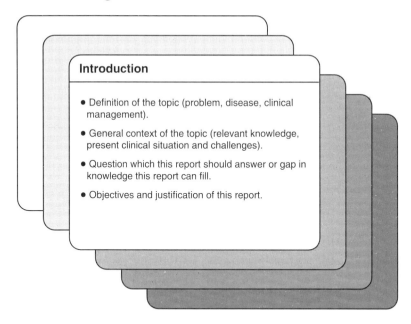

Introduction

- Definition of the topic (problem, disease, clinical management).

- General context of the topic (relevant knowledge, present clinical situation and challenges).

- Question which this report should answer or gap in knowledge this report can fill.

- Objectives and justification of this report.

3.1. Therefore, the value of the case report would depend partly on the documentation of the newness of the concepts advanced by the case or on the possible resulting modifications[36,37] of the currently accepted view of the problem.

The newness of a case report may be categorized in three ways:

1. A very ordinary situation may present unusual features or occur in an unusual setting. For example, *Salmonella typhi* osteomyelitis of the sternum was noted in an immunocompetent patient[38]. Elsewhere, *Salmonella* skull osteomyelitis was observed[39].

2. A rare or forgotten disease whose typical features have appeared in recent times in a specific setting (e.g. necrotizing fasciitis), thus leading to the improvement of diagnostic and timely treatment techniques.

3. Sometimes, although a certain disease has supposedly been eradicated, a few case reports in the literature can

be indicative of a new occurrence (endemicity, pandemic or epidemic).

Background and general context of the problem. A very succinct review of the problem tells the author whether or not a case[5] should be reported. An adequate literature search is also necessary for the author to know what clinical and paraclinical data should be gathered and studied.

Origins and motives of the report. Does the question under study relate to the risk assessment, diagnosis, treatment or prognosis? Does it lead to a better choice from a set of possible clinical decisions? A reader with less clinical experience will especially benefit from such explicit information.

Objectives and justification of the report. Specifying the expected results with regard to practice and/or research is probably the most difficult part of stating the relevance of the report. Despite this, the report's justification should always be present, either in this section or in the conclusion of the case report (with the addition made by the contribution to our knowledge of the problem).

3.2.2.3.4 Presentation of the case

This section is the core of the message. Table 3.5 summarizes its components.

A clinical case report is an exercise in **clinimetrics**[25,26], originally defined as *the measurement of clinical data*[25], or as *the field concerned with indexes, rating scales and other expressions used to describe or measure symptoms, physical signs or other distinctly clinical phenomena in clinical medicine*[40,41] *in view of clinical decision making*[26]. It is also an exercise in **hermeneutics**, i.e. in *interpretation and understanding*[42]. Now, the reader may be wondering: 'Do I do really need all this?'. The answer is, definitely. Although this process occurs most often unconsciously, it may also take place consciously in accordance with a certain structure.

Table 3.5

Components of the section presenting the case

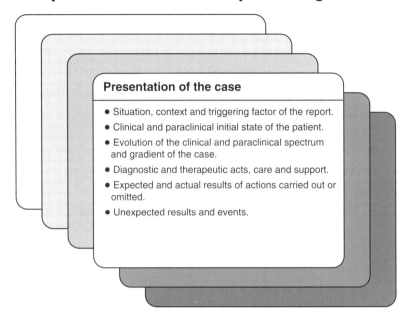

Presentation of the case

- Situation, context and triggering factor of the report.
- Clinical and paraclinical initial state of the patient.
- Evolution of the clinical and paraclinical spectrum and gradient of the case.
- Diagnostic and therapeutic acts, care and support.
- Expected and actual results of actions carried out or omitted.
- Unexpected results and events.

Thinking and evaluation in clinical epidemiology are based on the paradigm of a simple sequence: **initial state** → **action (manoeuvre)** → **subsequent state**[24,26,43] that is often tricky to apply, analyse and interpret in many actual situations. Clinical case presentations follow a pattern similar to the one illustrated in Figure 3.1.

1. **Antecedents and development of the case**. This includes family and personal history, exposure to aetiologic factors and incubation periods plus developments of the disease in its induction period.

2. The **initial state** is the moment at which the case report begins.

3. Clinical and paraclinical **manoeuvres** or **actions** follow. They may be diagnostic in nature such as serum ferritin measurement, magnetic resonance examination, neurological assessment etc. The probability of diagnosis before and after results (initial and subsequent state) is also reviewed. This may involve drug treatment or surgery with the expectation of a better subsequent state.

4. The **subsequent state** or **result (outcome)** is then outlined, i.e. better health, a cure, unexpected complications, adverse effects.

5. Finally, ensuing **clinical decisions** are presented with regards to the treatment, the improvement and expansion of the diagnostic work-up, and the precautions to be taken in similar cases and situations.

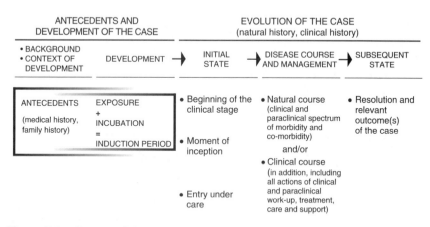

Figure 3.1 – Stages of the case

A case report should always contain these steps in an easy to understand format.

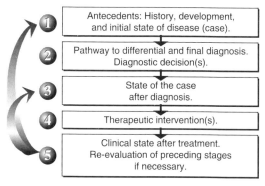

Figure 3.2 – Sequence of case management

The management of every case should follow the chronological sequence illustrated in Figure 3.2, with its alternate data collection, diagnostic process, treatment and effect and patient outcomes. One or several feedback loops within this cascading structure are common. A case report should also allow the reader to understand the steps that the author took to progress through such a sequence.

Existing situation in the field of interest and triggering factor of the report. The reader of a case report is not necessarily as familiar with the pathology and its clinical management as the author. A summary of current practices and acquired experience generally accepted in the field should be presented. The reason for the report should also be given.

For example, the challenges and highlights of a particular cancer treatment should be related first. Then, the unexpected adverse effects of treatment should be outlined (nausea, abnormal neurological, haematological and biochemical findings, haemorrhagic enterocolitis, anaphylactoid reactions etc.). The author of the report should then explain which expensive intensive care and critical outcomes in the patient should be avoided in the future, if a subsequent pharmaco-epidemiological study showed a cause–effect relationship between the treatment and its suspected undesirable effects, as suggested by the reported case.

Initial clinical and paraclinical state of the patient. Often, clinical data (interview, bedside examination and follow-up) and paraclinical data (microbiology, haematology, biochemistry, imaging techniques etc.) should be identified according to their role in the study and management of the pathology in question.

Patient characteristics. From a patient's history and physical examination, several *risk characteristics* can be identified. It should be specified which of these are *risk factors* (modifiable, such as smoking, physical inactivity or eating habits) and which are *risk markers* (non-modifiable, such as age, sex or race). Risk characteristics precede the onset of the disease, just as smoking precedes lung cancer. Other characteristics still present or appearing after the onset of disease are of equal interest as *prognostic characteristics*, sometimes modifiable (*prognostic factors*), sometimes not (*prognostic markers*). Certain readers consider these subtleties truly important for decision making. From such 'networks of risk or prognostic characteristics' causal associations may be hypothesized or quantified if their strength and specificity are known. Further details belong to the field of epidemiology[26].

Course of the disease. As a result of drawing a picture based on direct experience and patient charts, less experienced reporters tend to provide all details from a clinical and paraclinical follow-up in order to assure their peers and case report recipients that nothing was forgotten and that the work-up and follow-up were complete.

Only essential information for case understanding is necessary. By way of example, why should normal levels of uric acid in body fluids in an overwhelming and fast evolving necrotizing fasciitis be reported if this information is not essential for decision making?

Nevertheless, some protocol of case reporting will be required sooner or later, even if it is only going to be kept at hand. It is also useful if readers raise additional questions about the management of the case: How was the normality of an observation defined in operational terms? If an outlying observation was made, was it a statistical outlier, or was it a clinical outlier implying different medical or surgical management? Was it an abnormality in terms of appearance, localization, tissue composition, function, rhythm, frequency, direction, volume, speed etc.?

Were the exploratory surgical procedure (e.g. access of a fibre-optic catheter) and the surgery itself (e.g. laparoscopic cholecystectomy) normal?

Evolution of the case's clinical spectrum and gradient. It is not always easy to describe a clinical course. Challenges lie in the identification of the beginning stages of the case and of the disease occurrence, in the follow-up of the disease spectrum and gradient, in the study of the co-morbidity, and in knowing where to stop.

Beginning of the case. This point in time is sometimes quite difficult to determine. The period of exposure often overlaps or blends with the incubation period of the case. Smoking and chronic obstructive lung disease is an example of this. The starting point of the disease may be insidious and the first mild spells might go unnoticed. The disease does not necessarily begin at the moment when the physician sees his patient for the first time. The problem of exposure and incubation is somehow obviated by epidemiologists who consider the sum of both these periods as an *induction period of the disease*.

Disease spells. Is the first reported spell of a disease really the first? This is not simply an academic question. If the spell is really the first, the situation is one of disease incidence. Related characteristics are then interpreted as *risk factors*. Any subsequent spell is considered an expression of the *point prevalence* of the health problem and the related characteristics are interpreted as *prognostic factors*. Ensuing clinical decisions are quite different. Smoking is a powerful risk factor for lung cancer, but no one has yet demonstrated convincingly that stopping smoking after the diagnosis of lung cancer meaningfully improves the patient's prognosis.

Angina, epilepsy, migraine, arthritis, amyotrophic lateral sclerosis, Alzheimer's and Parkinson's disease all fall into the challenging category of clinimetrics and epidemiological interpretation of findings.

Needless to say, all spells should be observed and measured equally well in terms of their timing, duration and amplitude. Remissions that distinguish them from another should also be studied.

Cases as part of the clinical spectrum. The clinical spectrum, like the colour spectrum, primarily reflects the variability of the clinical picture and the extent of the disease. For example, the ocular, ulceroglandular or respiratory manifestations of tularaemia indicate that this disease has an extended clinical spectrum. The common cold, however, has a very limited clinical spectrum.

In many clinical reports, greater attention is paid to the paraclinical spectrum. Nevertheless, it should be noted that the clinical spectrum of signs and symptoms merits equal attention and a rigorously structured description.

Cases as part of the clinical gradient. Studying the gradient of a disease is synonymous with studying its severity. Non-apparent cases, flu-like cases, symptomatic hepatitis and fulminant hepatitis reflect an extended gradient of viral hepatitis. Again, the clinical gradient of the common cold is very limited.

Abnormal white blood cell counts, elevated C-reactive protein, bacteraemia or antibody levels indicate where to place the case on the *paraclinical gradient* of an inflammation process or immunological state.

Some disease manifestations may reflect both disease gradients and spectrums, such as an *in situ* growth, a metastatic spread and other components of cancer staging.

Contemporary medical literature abounds with the subject of *clinimetric indexes*. Some of these measure the spectrum and gradient of a single disease. Others do the same while covering a larger set of diseases and states, and yet others help to accurately categorize cases from among competing diagnoses. The *Glasgow Coma Scale,* the *Apgar Score,* the *Injury Severity Score (ISS),* the *APACHE system (Acute Physiology and Chronic Health Evaluation)* and the *Canadian Neurological Scale* are just a few examples.

These indexes should be carefully chosen and interpreted. Recommendations on how to do so have already been outlined in the clinical epidemiological literature[24,25,36] and are well beyond the scope of this reading.

Clinical and paraclinical data. Some states may be measured clinically or paraclinically. By way of example, physical examinations and haematological and biochemical findings may both be used to study malnutrition. Since space is generally limited in case reporting, clinical and paraclinical data should be chosen and reported as complementary parts of the case picture rather than as simple duplicate measurements of the same problem.

Co-morbidity. Depending on the problem under study, co-morbidity and its treatment may require a study and a presentation as precise as those of the disease of principal interest. This would help define the interactions between the former and the latter.

Diagnostic procedures, treatment intervention and care. In a clinical case report, the information being presented is almost always limited by time, if the communication is verbal, or by space, if the communication is presented in a journal. However, whenever the reader requires certain details in order to understand the risks, consequences, complications or successes in the management of the case, all clinical and para-clinical manoeuvres should be included in the clinical case study protocol and selectively presented in the report.

Expected as well as unexpected actions and events should be recorded. Intercurrent disorders and their treatment and the reasons for reporting them should also be outlined.

Outcome reporting focuses solely on the main outcome of interest. In other types of recording, *webs of consequences and outcomes* merit attention. For example, a psychiatrist might want to know not only about the drug abuser's sensorium, mood or cognitive functioning, but also about the patient's malnutrition, infections, marital and occupational problems, or delinquency. Is the reporting of these details necessary for a reasonable understanding of the case?

End-point of the case. Should the case report end at the moment of discharge from the hospital? Should events leading up to the admission and those occurring after the discharge be described in the case report? It is up to the case reporter to decide.

To summarize, an overall clinimetric description of the case should have qualities similar to those prized in writing and playing music[36]: movements creating a good component structure (disease stages), melody (construct validity, i.e. necessary components are like notes in a musical composition), harmony (clinical spectrum), dynamics (disease gradient) and rhythm (frequency, amplitude, and duration of disease spells and remissions).

In lieu of a conclusion, Table 3.6 outlines the clinical epidemiological challenges of case studies and reporting, including related precautions which should be known to the reporter and reader (and put into practice, if needed). *Obviously, all this information does not necessarily have to appear in the report. The author should nevertheless be ready to answer any questions raised regarding these matters.*

Table 3.6

'The Twelve Commandments': Clinimetric criteria, rules and precautionary measures relevant to clinical case reporting

Clinimetric criteria and challenges	Precautionary measures
• Past and present risk characteristics of the patient	*Make clear distinctions between characteristics which can be modified (risk factors) and characteristics that can't (risk markers). Choose only those characteristics that are relevant to the case under study.*
• Clinical data and information	*Treat clinical data (measurements) and information (interpretations) as separate entities. Choose only data and information that are relevant to the diagnosis.*
• Criteria for clinical data and information	*Clearly define the diagnostic criteria and make them operational. Clearly distinguish between normalcy and abnormality. Usually, conceptual criteria are insufficient.*
• Syndrome and disease; two different clinical entities	*A syndrome is a syndrome and a disease is a disease. A syndrome is a set of often dissimilar manifestations common to several causes, diagnostic entities and aetiologies known or unknown, that do not necessarily lead to treatment. A disease is an entity (contrary to the above) with a distinct manifestation, cause, and treatment.*
• Clear notions of clinical signs, symptoms, and other phenomena	*In different circumstances and situations, the same clinical manifestation can be observed and recorded either as a sign (objective) or as a symptom (subjective).*
• Appropriate preservation of hard and soft data	*An attempt should be made to harden soft data whenever possible and the method of hardening should be known (definition, direction, analogy, recording).*

Table 3.6 continued

● Description of disease course (evolution of the case)	*Specify whether the description is based on the disease spectrum (extent), the disease gradient (severity), or both. Choose the best clinimetric indexes for measuring disease severity whenever a directional disease measurement is necessary.*
● Follow-up and recording of information regarding clinical and paraclinical procedures	*Always note the frequency, intensity, quality, quantity, and hierarchic organization of diagnostic (serial or parallel testing) and therapeutic manoeuvres.*
● Evaluation of the effect of a diagnostic or therapeutic manoeuvre	*Determine and be aware of the internal and external validity of the diagnostic procedures used, as well as their impact on clinical decision-making (treatment or no treatment, further diagnostic work-up, etc.). If known, keep in mind the effectiveness of preventive and therapeutic interventions in terms of aetiological and prognostic fractions.*
● Actions chosen which have clinical consequences and which influence diagnostic or therapeutic results	*The treatment and outcome of a disease require the same rigorous definition and description as does the presentation of this disease, its progress, and management. Try to harden soft data whenever possible.*
● Prognostic characteristics	*Make clear distinctions between risk characteristics and prognostic characteristics. Specify the prognostic markers (non-modifiable) and the prognostic factors (modifiable).*
● Follow-up of co-morbid states and their treatment	*A rigorous and detailed study of co-morbidity and treatment of co-morbidity are necessary for a better understanding of differential diagnosis and preferred and adopted therapeutic decisions. The clinical and paraclinical management of all relevant morbid states should be equally well known, described, analysed, and explained.*
N.B. Other details and fundamental notions of clinimetrics can be found in textbooks devoted exclusively to clinical epidemiology and clinimetrics.	

3.2.2.3.5 Discussion and conclusion

In this section, the author analyses the findings from his case, presents his conclusions and gives his *recommendations* for future work. These points are usually covered at the three levels illustrated in Table 3.7.

Three important questions must be answered in the discussion and conclusion section:
- What conclusions can be drawn from all the data obtained from the case?
- How can a synthesis of the case be produced in view of the actual experience illustrated by the case?
- What future actions should be taken in the light of this experience?

Let us now comment on the three points described in Table 3.7.

Discussion of observations and results – possible consequences of the case experience. As in any other scientific paper, this is the only section where the author has the freedom to express criticisms, doubts and questions related to reported observations. Such a display of 'enlightened uncertainty' should help the reader understand the scope and range of the conclusions. Presenting confronting ideas in this section is also an option.

Any part of the case follow-up that might cause confusion or uncertainty, or require special attention, or represent an unusual observation should be highlighted.

Table 3.7

Components of the discussion and conclusion

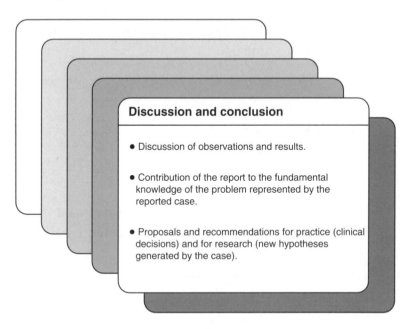

Discussion and conclusion

- Discussion of observations and results.

- Contribution of the report to the fundamental knowledge of the problem represented by the reported case.

- Proposals and recommendations for practice (clinical decisions) and for research (new hypotheses generated by the case).

There are two stages and levels of discussion based on the following questions:
- What is of interest solely in terms of this case report? This is a *within the case discussion*.
- What is of interest in a context larger than the case itself? This is a *beyond the case discussion*.

This section often takes up as much space as the case presentation itself, especially if the authors want (and they should):

- to compare their case with other cases of interest reported in the past;
- to specify the contribution of this case observation to the basic understanding of the disease or situation illustrated by the case.

Contribution of the case to the basic knowledge of the problem under study. What elements are helpful in understanding the problem and its fundamental mechanisms in terms of pathology, physiology, etc.? Can the pharmacodynamics of a given drug treatment be better understood? What are the morphologic or functional consequences of surgery? All these questions require answers.

What is 'new' and therefore worthy of discussion? Any phenomenon that has not yet been evaluated and elucidated in a given context as described in the case. For example, in the area of medical technology, any technology can be considered 'new' if its safety, efficacy and effectiveness, cost-effectiveness and cost–benefit ratios, and ethical, legal and social impact are unknown.

Proposals and recommendations for practice and research. To avoid unjustified generalizations, the analysis of a case must remain limited. The term **idiographic** analysis[37] was proposed for an analysis which is exclusive (or limited) to the case under study. This kind of analysis and its related recommendations also apply to the clinical case report.

However, special consideration should be given if the case is important for the basic sciences or for practical clinical purposes. Does the case represent a situation that requires more highly focused decisions? Is the case experience also important beyond the field of health sciences? Are there possible technological, social, cultural or economic implications?

Several obstacles stand in the way of tempting generalizations based on the single case, such as the obviously different demographic, biological and social characteristics of the patients, the severity of the cases, the co-morbidity and its treatment, the different treatments preferred by different physicians, the different care provided, the different migration of cases from one service or institution to another, and the admission policies.

For example, if a near fatal reaction is observed after a patient suffering from cardiac arrest is administered a sympathomimetic drug, its report alone does not warrant a contraindication right away. This observation should lead instead to a pharmaco-epidemiological study (e.g. case control study) which would demonstrate more clearly if such treatment in similar situations should be continued or rejected, given that the adverse effects have just been proved more effectively.

If differential diagnosis challenges are illustrated by the case, this should be a focus of the section. If difficult choices between competing treatments are reported, discussion and recommendations related to this problem should be included. Are rather unexpected complications and effects of treatment important? The reader does not necessarily know if they are, but should be told.

Do we always have to arrive at pragmatic conclusions? This depends. It is certainly important to present an unusual case and, if other similar cases appear, an epidemiological occurrence study of this phenomenon should be considered. Other cases are added to complete numerators, and community data provide denominators for rates of events of interest, which can be interpreted by epidemiological standards. Such uses may justify the multiplication of messages from various cases, ultimately leading to other pragmatic conclusions.

If prognostic or risk factors are of special interest in a case report, an adequate operational meaning should be given to them whenever necessary information is available. For example, a given treatment might be considered effective and a good prognostic factor. What do we know about the treatment's effectiveness beyond this case, or in other cases and situations? How might we reassess our treatment of the case in the light of such additional information?

Do we know something about the drug or surgery under study in terms of the following three points?

1. What is its **absolute risk reduction (ARR)**[44,45] power, i.e. the absolute arithmetic difference in the bad event rates (i.e. undesirable events such as non-responsiveness to treatment, complications, death etc.) of untreated and treated subjects ($Rate_{untreated}$ − $Rate_{treated}$)? This measure is known to epidemiologists as the *attributable risk (AR)* or *risk difference (RD)*[36,46]. It shows what the true rate of bad events in untreated individuals is, since these subjects did not benefit from the treatment under study.

2. What is its **relative risk reduction (RRR)**[44,45] power, i.e. the proportional reduction in rates of bad events between control and experimental participants in a trial? This measure is known to epidemiologists as the *aetiologic fraction (EF), attributable fraction (AF), attributable risk per cent (AR%),* or *protective efficacy rate (PER)*[36,46]. It shows what proportion of an overall rate of bad events in untreated subjects would be avoided if untreated subjects benefited from treatment: ($Rate_{untreated}$ − $Rate_{treated}$ / $Rate_{untreated}$) × 100. A result of 0% shows no protective efficacy or relative risk reduction. A result of 100% indicates absolute efficacy; i.e. treatment would prevent bad events in all untreated cases.

3. What does the **number needed to treat (NNT)**[44,45,47,48] mean for the use of this drug? This measure indicates the number of patients who need to be treated to achieve one additional favourable outcome. It is calculated as 1/ARR. If the clinician knows an individual patient's specific baseline risk (patient expected event rate) or **F**, an NTT for the specific patient may be calculated as NNT/F[45].

When the treatment increases the probability of a good event, an **absolute benefit increase (ABI)** instead of the ARR and a **relative benefit increase (RBI)** instead of the RRR may be estimated. When the treatment increases the probability of a bad event, **absolute risk increase (ARI), relative risk increase (RRI)**, and the **number needed to harm (NNH)** may be also calculated[45].

When the role of a particular prognostic or risk factor in a clinical case report is discussed, this may be done by creating an analogy related to other situations and experiences reported in the literature in the above-mentioned terms.

If a treatment works in a single case, we can only guess that an outcome of interest is really due to this particular treatment. This should be confirmed by an analogous experience, extraneous to the case under study.

In summary, we might be interested if the diagnostic and therapeutic decisions surrounding the case are evidence-based. Also, this approach might indicate which interesting risk or prognostic factors in a single case should be evaluated by a more powerful method of gathering evidence in the future.

Hence, choosing and presenting valid clinical and paraclinical data in a single case report is fundamental as a basis for evidence. Critically discussing the case from the point of view of the evidence in the literature and making conclusions based on the same principles progressively leads us to **evidence-based clinical case reporting**.

3.2.2.3.6 References

Like the other sections of the case report, the references should also be limited in terms of contents. The number of references should be reduced in such a way that only those references required for the basic understanding of the message should be cited. One reference covering the specific problem under study and another more general source are sometimes an acceptable minimum. One review article and some other references necessary to understand the contradictions between the findings from this case and those from other reports are also suitable.

Table 3.8 shows that references should cover three areas: the disease or problem under study, the clinical actions, and the decisions to be considered after this case experience.

One of the frequent problems with references is that they often cover just one of the three categories stated. For example, the reader may want to further study the disease, but he may not know where to start expanding his knowledge of

Table 3.8

The areas references should cover

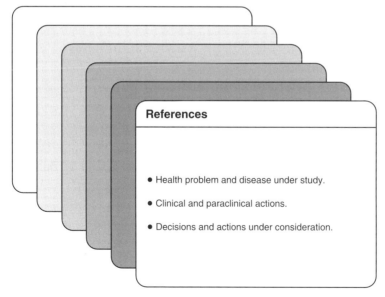

References

- Health problem and disease under study.

- Clinical and paraclinical actions.

- Decisions and actions under consideration.

clinical manoeuvres or complex decision points. Everything depends on the focus of the case report.

3.2.2.3.7 Recapitulation and synthesis

Once all the above-mentioned stages are complete, the organization and structure of a case report should follow the guidelines in Table 3.9.

The art of clinical case reporting relies on the author's wise selection of data and information, and on his ability to reduce and compress them within an allotted space. He must carefully balance the length of each section without losing the essential content of the message.

The underlying process is similar to the task assigned to Hermes in Greek mythology, whereby he had to interpret the gods' messages and make them understandable for humans. In other words, Hermes had to transform something from unintelligibility to understanding[49]. This is also the purpose of a case report: to draw out the most essential information from a maze of clinical data.

3.2.2.4 Form of a clinical case report

How can one respect all the rules of reporting and still fit all five sections of Table 3.9 into the required formats of the medical literature?

There are two prevailing attitudes in the area of clinical case reporting. On the one hand, periodicals prefer to publish only truly exceptional cases. If this is the situation, they are ready to offer as much space as needed. For example, the *Journal of the American Medical Association (JAMA)* has in the past devoted two full pages to a valuable clinical case report. On the other hand, case reports that are a more regular component of an issue are usually limited to between 500[50] and 1500 words (seven or eight double-spaced pages for some[51-53], four[54,55], six[18] or ten[56] pages for others). In addition, not more than three authors' names, three to six key

Table 3.9

Architecture of the clinical case report

Summary*	• Motives and reasons for the report.
	  'Why are we reporting this?'
	• Background to the problem.
	  'In what context?'
	• Highlights of the report.
	  'What have we found?'
	• Conclusions.
	  'What does this mean?'

Introduction	• Definition of the topic (problem, disease, clinical activity). • General context of the topic (relevant knowledge, present clinical situation and challenges). • A question that this report should answer or a gap in knowledge this report can fill. • Objectives and justification of this report.
Presentation of the case	• Situation, context and triggering factor of the report. • Clinical and paraclinical initial state of the patient. • Evolution of the clinical and paraclinical spectrum and gradient of the case. • Diagnostic and therapeutic acts, care and support. • Expected and actual results of actions carried out or omitted. • Unexpected results and events.
Discussion and conclusion	• Discussion of observations and results. • Contribution of the report to the fundamental knowledge of the problem represented by the reported case. • Proposals and recommendations for practice (clinical decisions) and for research (new hypotheses generated by the case).
References	• Health problem and disease under study. • Clinical and paraclinical actions. • Decisions and actions under consideration.

*NB A summary of a clinical case report is not required by all medical journals.

words, two figures or tables, and five[51] to fifteen[53] references can be included.

According to the editors of *The Canadian Medical Association Journal* [55], *'a case report should comprise a title page, a 1000 word summary, an introduction, the case report, comments and references, and normally not exceed 800 words* (four double-spaced pages), *excluding the title page and references'*.

The case report must be written in simple, precise and concise language. Jargon, clichés, professional 'barbarisms', illogicalities and fashionable buzz-words, as well as 'lumpiness' of terms, should be avoided[10,51].

Let us also remember that presentations of rounds must adhere to the same rules[57].

Ethics and confidentiality must underlie any case report. A case titled 'Hip replacement at the Vatican for the leader of one of the world's major religions' demonstrates the author's lack of mastery of both above-mentioned virtues.

3.3 GENERAL AND SPECIFIC EXPECTATIONS OF MEDICAL JOURNALS FOR CASE REPORTS

Clinical case reports should conform to the same rules as other medical articles[58–60]. Checklists[61] can help achieve this goal.

As paradoxical as it may seem, a case report should address and explain similar issues to those encountered in review articles[62]. These issues generally relate to the reasons for the report, the sources of information, the rules governing the selection and presentation of information, the valid materials, the systematic gathering of data, the review of possible limitations and inconsistencies, the summary, and the recommendations.

Table 3.10 summarizes the *Canadian Medical Association Journal* expectations with regard to case reports[17]. The reader

Table 3.10

CMAJ expectations with regard to case reports (reproduced with kind permission of *The Canadian Medical Association Journal*)

Summary
- Does the summary give a succinct description of the case and its implications?

Introduction
- Is the case new or uncommon, with important public health implications?
- Is the rationale for reporting the case adequately explained?
- Has an adequate review of the literature been done?

Description of the case
- Is the case described briefly but comprehensively?
- Is the case described clearly?
- Are the results of investigations and treatments described adequately, including doses, schedule and duration of treatment?
- Are the results of less common laboratory investigations accompanied by normal values?

Comments
- Is the evidence to support the author's diagnosis presented?
- Are other plausible explanations considered and refuted?
- Are the implications and relevance of the case discussed?
- Is the evidence to support the author's recommendations presented adequately?
- Does the author indicate directions for future investigation or management of similar cases?

should also refer to Tables 3.1 and 3.9 as additional guides in preparing his own clinical case presentation.

3.4 IN CONCLUSION

A clinical case report is an endeavour as serious as any other article or scientific paper in medicine. Its content and form should live up to the expectations that are now clearly stated in many medical journals. Let us respect these expectations.

A last word of advice to future medical casuists. The required brevity of clinical case reports is not an excuse for hiding erroneous or missing data and findings. Good data and information, often beyond the scope of the case report, should always be at hand. In fact, they are essential[47,63,64].

The ability to produce a concise, to the point presentation is mainly an acquired, learned and mastered skill.

Independent of the case report's length, it should result form a serious in-depth qualitative analysis.

Psychiatric and psychoanalytic work-ups and evaluations[65], patient follow-ups and case presentations are hermeneutic and phenomenologic processes[66] *par excellence*.

A clinical casuist observes, measures, records and establishes a **theory**; i.e. he selects an *interrelated set of constructs (or variables) formed into propositions or hypotheses that specify the relationship between variables*[66]. These variables must be carefully chosen. Not all of them are necessary, depending on the purpose of the study. If aetiology is the focus of attention, variables constituting a possible network of causes must be evaluated. If prognosis is the main subject, variables constituting possible networks of consequences should be taken into consideration. If possible sources of confusion such as co-morbidity, treatment for co-morbidity etc. are important, they should be included in the observation or interpretation of the case.

Consequently, as already mentioned in Section 3.2.2.3.4, a case report is methodologically based on a blend of qualitative research from its conception[67] to its presentation[68], and on clinimetrics[24-26]. It also supports recent endeavours to integrate qualitative and quantitative research methods[69].

Certainly, clinical case reporting is the weakest step in the lines of medical evidence. However, it is also medicine's first line of contact, so it is that much more important for the case report to be right. If the report is based on a disorganized and unfocused approach, and poor and incomplete observations, recordings, analyses and interpretations, it is of little value. The challenge is to create a report in such a way that building stronger evidence based on the case report is worth the effort.

A clinical case report cannot extend beyond its own purposes, namely to explore, describe, explain and raise questions. This fact is even more important given that many new challenges have originated from case reports: genetic disorders and mapping, organ transplantation, microsurgery, *in vitro* fertilization, cloning, genetically engineered vaccines and drugs etc. Stronger evidence of the *raison d'être* of such new technologies has usually followed.

Even a case report, despite its own inherent limitations, must:
- have an acceptable focus (problem and question);
- be sound;
- be well structured;
- be based on a known process; and
- be specific and clear for the reader, in terms of its message.

Moreover, any researcher can surpass the efforts of a clinical case reporter by building on his or her own case experience by planning and undertaking a study of numerous cases in a case series report or in an occurrence study (a descriptive study in epidemiological terms) or by comparing two or more groups in explanatory research.

If the next lines of evidence in medicine rely on a poorly selected index case, the whole process of

> producing evidence is at high risk of being
> diverted along the wrong path.

It is hoped that this chapter illustrates at least in part that the whole exercise of clinical case reporting extends far beyond an intern's wish to 'just survive' his case presentation[70].

Does the reader now need to temporarily escape from this Cartesian approach to clinical case reporting? A funnier, but eloquent, review of case report thinking in Bennett's book[71] is an uplifting remedy for the reader's probable state of mind. When the reader does pick up this text again, Chapter 4 will present an example of a good clinical case report. Chapter 5 will then show how to report a set of cases as a clinical case series report.

REFERENCES

1. Yurchak PM. A guide to medical case presentations. *Res Staff Phys,*1981 (September);**27**:109–15.
2. Zack BG. A guide to pediatric case presentations. *Res Staff Phys,*1982 (November);**28**:71–80.
3. Pickell GC. The oral case presentation. *Res Staff Phys,*1987(June);**33**:141–5.
4. Locante J. The anatomy of a perfect case presentation: A – Z. *J Oral Implantol,* 1996;**XXII**:65–7.
5. Huth EJ. The case report (Chapter 6, pp. 58–63) and The review and the case-series analysis (Chapter 7, pp. 64–68) in: *How to Write and Publish Papers in the Medical Sciences.* Philadelphia: iSi Press,1982. (See also 2nd Edition. Baltimore: Wiiliams & Wilkins,1990.)
6. Simpson RJ Jr, Griggs TR. Case reports and medical progress. *Persp Biol Med,*1985 (Spring);**28**:402–6.
7. Coates M-M. Writing for publication: Case reports. *J Hum Lact,*1992;**8**:23–6.
8. Freeman TR. The patient–centred case presentation. *Fam Pract,*1994;**11**:164–70.
9. DeBakey L, DeBakey S. The case report. I. Guidelines for preparation. *Int J Cardiol,*1983;**4**:357–64.
10. DeBakey L, DeBakey S. The case report. II. Style and form. *Int J Cardiol,* 1984;**6**:247–54.

11. Kroenke K. The case presentation. Stumbling blocks and stepping stones. *Am J Med*,1985;**79**:605–8.
12. Mott T Jr. Guidelines for writing case reports for the hypnosis literature. *Am J Clin Hypnosis*,1996;**29**:1–6.
13. Lilleyman JS. How to write a scientific paper – rough guide to getting published. *Arch Dis Childhood*,1995;**72**:268–270.
14. Levin M. How to write a case report. *Res Staff Phys*,1998;**44**(1):60–3.
15. Squires BP. Case reports: What editors want from authors and peer reviewers. *Can Med Assoc J*,1989;**141**:379–80.
16. Squires BP, Elmslie TJ. Reports of case series: What editors expect from authors and peer reviewers. *Can Med Assoc J*,1990;**142**:1205–6.
17. Huston P. Squires BP. Case reports: Information for authors and peer reviewers. *Can Med Assoc J*,1996;**154**:43–4.
18. Riesenberg DE. Case reports in the medical literature. *JAMA*,1986;**255**:2067.
19. *Landmark Articles in Medicine*. Edited by HS Meyer and GD Lundberg. Chicago: American Medical Association, 1985.
20. Levine P, Stetson RE. An unusual case of intra-group agglutination. *JAMA*,1939;**113**:126–7. (Reprinted in: *JAMA*,1984;**251**:1316–7.)
21. Levine P, Newark NJ, Burnham L, Englewood NJ, Katzin EM, Vogel P. The role of isoimmunization in the pathogenesis of erythroblastosis fetalis. *Am J Obstet Gynecol*,1941;**42**:925–37.
22. Greenwalt TJ. Rh isoimmunization. The importance of a critical case study. *JAMA*,1984;**251**:1318–20.
23. DeBakey L, DeBakey S. The case report. I. Guidelines for preparation. *Int J Cardiol*,1983;**4**:357–64.
24. Feinstein AR. *Clinical Epidemiology. The Architecture of Clinical Research*. Philadelphia: WB Saunders, 1985 (p. 241).
25. Feinstein AR. *Clinimetrics*. New Haven and London: Yale University Press, 1987.
26. Jenicek M. *Epidemiology. The Logic of Modern Medicine*. Montreal: Epimed International, 1995.
27. Kahn CR. Picking a research problem. The critical decision. *N Engl J Med*,1994;**330**:1530–3.
28. *How to Conduct a Cochrane Systematic Review*, 3rd Edition. Edited by CM Mulrow and A Oxman. San Antonio: San Antonio Cochrane Center, 1996.
29. Counsell C. Formulating questions and locating primary studies for inclusion in systematic reviews. *Ann Intern Med* 1997;127:380–7.
30. Yackee J, Lipson A, Wasserman AG. Electrocardiographic changes suggestive of cardiac ischemia in a patient with esophageal food impaction. 'A case that's hard to swallow'. *JAMA*,1986;**255**:2065–6.
31. Haynes RB, Mulrow CD, Huth EJ, Altman DG, Gardner MJ. More informative abstracts revisited. *Ann Intern Med*,1990;**113**:69–76
32. Mulrow CD, Thacker SB, Pugh JA. A proposal for more informative abstracts of review articles. *Ann Intern Med*,1988;**108**:613–5

33. *Ad Hoc* Working Group for Critical Appraisal of the Medical Literature (Haynes RB *et al.*). A proposal for more informative abstracts of clinical articles. *Ann Intern Med*,1987;**106**:598–604

34. Richtsmeier WJ. Case report. *Arch Otolaryngol Head Neck Surg*,1993;**119**:926.

35. Nahum AM. The case report: 'Pot boiler' or scientific literature? *Head Neck Surg*,1979;**1**:291–2.

36. Jenicek M. Making clinical pictures or describing disease in an individual,Section 5.2, pp. 125–136 in: *Epidemiology. The Logic of Modern Medicine*. Montreal: Epimed International, 1995.

37. Barr JE. Research and writing basics: Elements of the case study. *Ostomy/ Wound Management*,1995;**41**(1):18,20,21.

38. Weiss K, Laverdière M. *Salmonella typhi* osteomyelitis in an immunocompetent patient. *Clin Microbiol Newsletter*,1995;**17**:35–6.

39. Kamarulzaman A, Briggs RJS, Fabinyi G, Richards MJ. Skull osteomyelitis due to *Salmonella* species: Two case reports and review. *Clin Infect Dis*,1996; **22**: 638–41.

40. *A Dictionary of Epidemiology*, 3rd Edition. Edited by JM Last. Oxford and New York: Oxford University Press, 1995.

41. Feinstein AR. *Clinimetrics*. New Haven: Yale University Press, 1987.

42. Rothe JP. *Qualitative Research. A Practical Guide*. Heidelberg and Toronto: RCI/ PDE Publications, 1993.

43. Feinstein AR. *Clinical Biostatistics*. St Louis: CV Mosby, 1977.

44. Sackett DL, Richardson WS, Rosenberg W, Haynes RB. Evidence-based medicine. How to practice and teach EBM. New York: Churchill Livingstone, 1997.

45. Anon. Evidence-based medicine. Glossary. *Evidence-Based Med*,1999 (May/ June); **4**: inside back cover.

46. Jenicek M. Epidemiology, evidence-based medicine, and evidence-based public health. *J Epidemiol*,1997;**7**:187–97.

47. Laupacis A, Sackett DL, Roberts RS. An assessment of clinically useful measures of theconsequences of treatment. *N Engl J Med*,1988;**318**:1728–33.

48. Chatellier G, Zapletal E, Lemaître D, Ménard J, Degoulet P. The number needed to treat: a clinically useful nomogram in its proper context. *BMJ*,1996;**312**:426–9.

49. Addison RB. Grounded hermeneutic research, pp. 110–124 in: *Doing Qualitative Research*. Edited by BJ Crabtree and WL Miller. Newbury Park: Sage Publications, 1992.

50. Information for authors. *Intern J Cardiol*,1995;**51**:309–10.

51. Soffer A. Case reports in the Archives of Internal Medicine. *Arch Intern Med*,1976;**136**:1090.

52. Mott T Jr. Guidelines for writing case reports for the hypnosis literature. *Am J Clin Hypnosis*,1986;**29**:1–6.

53. Information for contributors. *Urology*,1994;**44**:153–N.

54. Information for authors. *Surgery*, 1994;**115**:7A–8A.

55. Instructions for authors. *Can Med Assoc J*,1995,**152**:55–7.

56. Instructions for contributors. *Am J Respir Crit Care Med*,1995,**152**:1/95–3/95.

57. Winter R. Edit a staff round. *Br Med J*,1991;**303**:1258–9.

58. Lilleyman JS. How to write a scientific paper – a rough guide to getting published. *Arch Dis Childhood*,1995;**72**:268–70.

59. Skelton J. Analysis and structure of original research papers: an aid to writing original papers for publication. *Br J Gen Practice*,1994;**44**:455–9.

60. International Committee of Medical Journal Editors. Uniform requirements for manuscripts submitted to biomedical journals. *Ann Intern Med*, 1997;**126**:36–47.

61. DuRant RH. Checklist for the evaluation of research articles. *J Adolesc Health*, 1994;**15**:4–8.

62. Milne R, Chambers L. Assessing the scientific quality of review articles. *J Epidemiol Comm Med*,1993;**47**:169–70.

63. Weed LL. Medical records that guide and teach. *N Engl J Med*, 1968;**278**:593–600. (Concluded in: *N Engl J Med*,1968;**278**:652–7.)

64. Koran LM. The reliability of clinical methods, data and judgements. *N Engl J Med*,1975;**293**:642–6. (Continued in: *N Engl J Med*,1975;**293**: 695–701.)

65. Shea SC. *Psychiatric Interviewing. The Art of Understanding.* 2nd Edition. Philadelphia: WB Saunders, 1998.

66. Creswell JW. *Research Design. Qualitative & Quantitative Approaches.* Thousand Oaks: Sage Publications, 1994.

67. Kuckelman Cobb A., Nelson Hagemaster J. Ten criteria for evaluating qualitative research proposals. *J Nurs Educ*, 1987;**26**:138–43.

68. Forchuk C, Roberts J. How to critique qualitative research articles. *Can J Nurs Res*, 1993;**25**(4):47–55.

69. Stange KC, Miller WL, Crabtree BF, O'Connor PJ, Zyzanski SJ. Multimethod research: Approaches for integrating qualitative and quantitative methods. *J Gen Intern Med*,1994;**9**:278–82.

70. Bennett HJ. How to survive a case presentation. *Chest*, 1985;**88**:292–4.

71. Case reports, pp. 117–135 in: *The Best of Medical Humor. A Collection of Articles, Essays, Poetry, and Letters Published in the Medical Literature,*2nd Edition. Compiled and Edited by HJ Bennett. Philadelphia: Hanley & Belfus, Inc., 1997.

Annotated example of a clinical case report

*Is there really an example
of a good clinical single
case report that meets the
stringent criteria
mentioned above?
Certainly, we have found
at least one!*

Annotated example of a clinical case report

4.1 INTRODUCTION TO THE EXAMPLE

The principles and requirements of a good clinical case report as reviewed in Chapter 3 should be followed in all clinical case reports in the literature. In reality, many reports are excellent and many are not. From the former pool, we have chosen one[1].

Yackee, Lipson and Wasserman have reported electrocardiographic changes as a challenge of differential diagnosis between an oesophageal impaction and coronary ischaemia[1]. Their case report elicited two important editorial comments.

Riesenberg[2] discusses a framework of good editorial policies regarding clinical case reports. Swirin and Hueter[3] analyse differential diagnostic challenges in situations where both cardiovascular and digestive disorders must be considered. To save space, these comments are not reproduced here. The reader is encouraged to find them in the literature[2,3] as suggested additional reading.

The pages that follow are an intact reproduction (with permission) of Yackee et al.'s report. In our own step by step annotations, places where good elements of a clinical case report have been put into practice have been underlined. We have also mentioned some missing points and possible areas of improvement. This, however, should not devalue in any way the authors' endeavour.

This clinical case report is worthy of attention for several reasons. First, the authors present an unexpected event. Then, they compare it to cases where electrocardiographic abnormalities were observed in conjunction with other digestive disorders. Finally, the authors discuss the practical implications for clinical decisions in similar situations. Obviously, this report has a well-structured, logical sequence that is easy to follow and grasp.

Relevant elements of the report are indicated by a graphic symbol (little square) corresponding to an identical sideline pictogram and related comments.

There should be enough page space to allow the reader to add his or her own comments and points to ponder. This interactive reading is strongly encouraged. Our comments are neither complete nor a reflection of the absolute truth. Work with this example while reading it!

Reminder: The case report reproduced here is a methodological example only. It is not in any way indicative of clinical diagnostic, therapeutic or prognostic decisions, which may change in time with increasing knowledge in the field under consideration.

4.2 ANNOTATED CASE REPORT

Electrocardiographic Changes Suggestive
of Cardiac Ischemia in a Patient With
Esophageal Food Impaction.

'A Case That's Hard to Swallow'*

John Yackee, MD, Ace Lipson, MD,
Allan G. Wasserman, MD**

□Many patients presenting with chest pain initially believed to be cardiac in etiology may, in fact, have esophageal disease as an alternative or additional cause of their complaints. □Although electrocardiographic (ECG) repolarization abnormalities have been well described in gastroenterologic processes, such as pancreatitis and cholecystitis[1], and have been reported in esophageal disease[2–9], the ECG is generally thought to be a reliable means of distinguishing between esophageal and cardiac pain[10,11]. □In this report we describe a patient who developed ECG changes suggestive of cardiac ischemia secondary to esophageal food impaction.

Report of a Case
□A 57-year-old woman was admitted to the George Washington University Medical Center, Washington, DC, with a

□The clinical relevance of the topic is highlighted.

□References to the literature summarize the overall knowledge of the problem as well as prevailing practices and decisions in similar circumstances.

□Objectives of this case report are stated here. (why the case is reported)

□The patient's medical history is outlined as it relates mainly to cardiology and gastroenterology, the areas of focus of this case's differential diagnosis. The patient's 'initial state' (at admission) is also reported.

*Reprinted from *JAMA* (*Journal of the American Medical Association*), 18 April 1986; 255:2065–6. (Copyright 1986, American Medical Association. Reproduced with permission.)
**From the Division of Cardiology, Department of Medicine, George Washington University Medical Center, Washington, DC.

chief complaint of chest discomfort and dysphagia. She had a history of intermittent epigastric and substernal burning discomfort with meals, alleviated by antacids, and in 1979 was found to have a hiatal hernia with reflux by an upper gastrointestinal (GI) tract X-ray series. For three to four years, she has noted intermittent dysphagia, with a sensation of solid foods 'sticking' in her throat. A repeated upper GI tract X-ray series 11 months before admission revealed the hiatal hernia with an area of mucosal irregularity in the distal esophagus, consistent with reflux esophagitis. She was managed with elevation of the head of her bed, antacids, thorough chewing of foods, and avoidance of caffeine and alcohol, with improvement. There was no history of exertional chest pain. On the day of admission, while eating steak at a dinner party after drinking several cocktails, she had the sensation of a piece of meat sticking in her throat, associated with subxyphoid and epigastric discomfort with a mild nausea and hiccoughs. She was unable to swallow liquids, regurgitating even her own oral secretions.

□Co-morbidity and treatment for co-morbidity are mentioned.

□Medical history included hypertension, manic-affective disorder, degenerative joint disease, aspiration pneumonia following a drug overdose, and excision of benign thyroid and breast nodules. Her medications included atenolol, 75 mg/day, and antacids.

Physical examination revealed a calm, moderately obese woman regurgitating her saliva into a cup. Blood pressure was 120/70 mmHg; heart rate 84 beats per minute and regular, respirations 18/min and unlabored. The cardiac and abdominal findings were unremarkable. Laboratory values were remarkable for a potassium level of 3.2 mEq/l. Chest X-ray film was normal. The ECG taken on admission revealed ST-segment depression and T-wave inversion in the inferior and anterolateral leads (Figure 1), which had not been evident on previous tracings. A barium swallow examination showed an irregularly contoured, intraluminal, esophageal filling defect proximal to the gastroesophageal junction (Figure 2).

Sublingual nitroglycerin and intravenous glucagon were administred, without resolution of symptoms.

Results of a clinical and paraclinical work-up at admission are presented as the patient's 'initial state' (e.g. ECG, radiography – see Figures 1 and 2).

Treatment according to the working diagnosis is defined and delivered (in order to exclude coronary co-morbidity), followed by the treatment (therapeutic manoeuvre) of the main problem under consideration (disimpaction of the oesophagus).

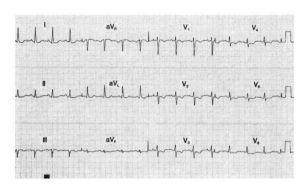

Figure 1 – Twelve-lead electrocardiogram during esophageal meat impaction. Note interior and anferolateral ST-segment depression and T-wave inversion

Esophagoscopy revealed a large piece of meat impacted in the distal esophagus at 35 cm. After administration of sublingual nifedipine, the meat bolus was

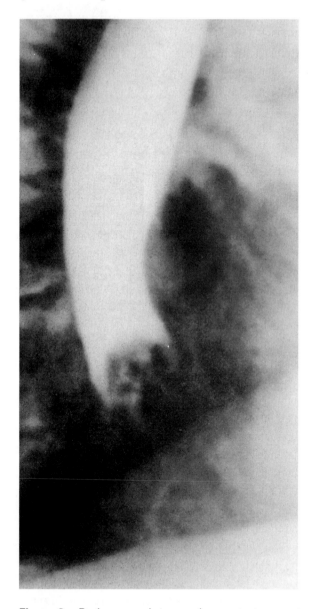

Figure 2 – Barium esophagram demonstrates meat bolus impacted in distal esophagus

CLINICAL CASE REPORTING

mechanically advanced into the stomach. Mucosal edema and superficial ulceration at the site of impaction, and a hiatal hernia distal to the site of impaction, were noted. There was no evidence of an esophageal stricture. An ECG obtained soon after endoscopic disimpaction showed marked improvement of the ST-T-wave changes (Figure 3).

☐The 'subsequent state' of the patient is reported and confirmed by the para-clinical follow-up (ECG after disimpaction).

Comment

☐ST-wave abnormalities have been reported in association with hiatal hernia,[12] esophageal spasm,[2–5,7–9] epiphrenic esophageal diverticulum,[6] and experimental mechanical distention of the esophagus.[10] However, we are aware of no other reported cases of ST-T-wave changes attributable to esophageal food impaction. This patient presented with chest discomfort and ECG changes consistent with cardiac ischemia. Although the history of previous esophageal symptoms, the subjective sensation of food

☐Challenges of the differential diagnosis from the cardiology and gastro-enterology points of view are given here with special attention to the ECG findings.

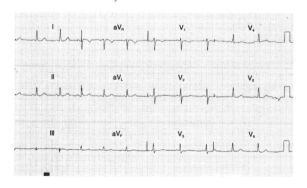

Figure 3 – Repeated 12-lead electrocardiogram after esophageal disimpaction. Note resolution of previous ST-T-wave changes

lodging in the throat, and the upper GI tract X-ray and endoscopic findings make the esophageal etiology of her chest discomfort readily apparent, this case has important implications regarding the reliability of ECG changes in differentiating cardiac and esophageal causes of chest pain. This differentiation can otherwise be difficult since historical features and the character of chest pain in each can frequently be indistinguishable.[2] Both can be alleviated by similar therapy, such as nitrates or calcium blockers, or precipitated by provocation testing, such as with ergonovine or methacholine chloride,[5,8] and both processes can occur in the same patient.[4,7]

□The authors justify and specify their final diagnosis.

□Explanations for this patient's ECG findings other than esophageal disease seem unlikely. Since our patient did not undergo cardiac catheterization with ergonovine provocation, a cardiac etiology for the ECG changes, such as ischemia precipitated by the esophageal impaction, cannot be completely excluded. However, the lack of previous cardiac history and the prompt resolution of the changes with esophageal disimpaction argue strongly against such a role of cardiac ischemia. The patient also has recently had normal results on a treadmill exercise test.

□A differential diagnosis is discussed by exclusion on the basis of the patient's history. It is important to note that the patient's past manic-depressive disorder has not been omitted either.

□Similarly, cardiac ischemia precipitated by coronary spasm as one component of a generalized disorder of smooth muscle, including a simultaneous eso-

CLINICAL CASE REPORTING

phageal motility disorder, seems unlikely.[7,8] The patient's history and ECG changes do not suggest coronary spasm.

Repolarization abnormalities have been reported in hiatal hernia[11] postprandially, with hypokalemia, anxiety, and psychiatric disorders, each of which might be invoked as playing a role in the present case. However, none of these is consistent with the patient's presentation, in particular, the transient appearance of the changes coincident with the development and resolution of the food impaction.

□The mechanism by which esophageal meat impaction might cause ST-T-wave changes is speculative. A vagal mechanism for similar changes observed in gastroenterologic conditions, such as pancreatitis and cholecystitis, has been proposed. Since gag is a powerful vagal stimulus, it would not be surprising that esophageal obstruction might provide such a stimulus. A mechanical shift of the T-wave axis secondary to a direct effect seems unlikely given only a minor change in the QRS axis. It is also difficult to conceive how a direct mechanical effect due to the proximity of the esophagus to the heart might induce ventricular repolarization abnormalities.

□Underlying possible pathological/ physiological mechanisms are discussed, which is proof that they have not been forgotten.

□In conclusion, a patient with underlying esophageal disease and no evidence of coronary artery disease developed marked ST-T-wave changes with esopha-

□A summary of the pathological/ physiological findings is presented.

geal food impaction, resolving promptly with disimpaction.

☐The originality of the observations and findings as a justification for this case report is stressed.

☐We believe this is the first reported case of ST-T-wave changes due to clinical esophageal impaction and speculate an autonomic esophageal–cardiac interplay as the mechanism. ☐This case provides additional evidence that the ECG is not always a reliable means of differentiating chest pain of cardiac and esophageal etiologies.

☐Decisions are proposed for similar situations in clinical practice.

☐References

1. Schamroth L (ed); *The Electrocardiology of Coronary Artery Disease*, ed 2. London, Blackwell Scientific Publications Ltd, 1984.
2. Davies HA, Rhodes J: How often does the gut cause angina pain? *Acta Med Scand* 1981;209(suppl 644):62–65.
3. Dart AM, Davies AH, Loundes RH *et al.*: Oesophageal spasm and angina: Diagnostic value of ergometrine (ergonovine) provocation. *Eur Heart J* 1980;1:91–95.
4. Edeiken J: Angina pectoris and spasm of the cardia with pain of anginal distribution on swallowing. *JAMA* 1939;112:2272–2274.
5. Davis AH,Kaye MD, Rhodes J *et al.*: Diagnosis of oesophageal spasm by ergometrine provocation. *Gut* 1982; 23:89–97.
6. Julian DG: Epiphrenic oesophageal diverticulum with cardiac pain. *Lancet* 1953;2:915–916.
7. Lee MG, Sullivan SN, Watson WC, et al: Chest pain: Esophageal, cardiac or both? *J Am Coll Cardiol* 1985;80:320–324.
8. Eastwood GL, Weiner BH, Dickerson WJ *et al.*: Use of ergonovine to identify esophageal spasm in patients with chest pain. *Ann Intern Med* 1981;94:768–771.

☐The references focus mainly on the challenge of differential diagnosis in similar situations of cardiac or digestive chest pain.

9. Davies HA, Jones DB, Rhodes J: Esophageal angina as a cause of chest pain. *JAMA* 1982;248:2274–2278.
10. Baylis JH, Kauntze R, Trounce JR: Observations on distension of the lower esophagus. *Q J Med* 1955;94:143–153.
11. Kramer P, Hollander W: Comparison of experimental esophageal pain with clinical pain of angina pectoris and esophageal disease, *Gastroenterology* 1955;29:719–743.
12. Delmonico JE Jr, Black A, Geusini GG: Diaphragmatic hiatal hernia and angina pectoris. *Dis Chest* 1968;53:309–315.

4.3 CONCLUDING COMMENTS

This 'straight to the point' report from the Washington University authors led to other considerations in the two above-mentioned editorial comments[2,3].

Usually, clinical reports from North America focus on decisions and actions. In contrast, case reports from Europe generally stress the understanding of the underlying physiological and pathological mechanisms of the case and clinical problem under study.

In fact, it is the author's duty to highlight the uniqueness of the case and its relevance from the point of view of the basic sciences (e.g. anatomy, physiology, pathology, pharmacodynamics), as well as its usefulness for diagnostic and therapeutic clinical decisions and for prognosis.

Meeting all these requirements remains a great challenge, especially if the message has to be conveyed briefly to a somnolent colleague after a weekend, to a confused and disoriented freshman, or to an overburdened senior staff person.

Clinical case reports must become more focused and structured in order to facilitate the reader's understanding. Since the author is responsible for the quality of his or her report, he or she may want to refer back to Chapter 3 of this book, which should serve as a useful tune-up checklist for the emission (author) and the reception (readers).

Last but not least, to fulfil the clinical case report's purpose and objectives, more than the one or two customary paragraphs may be needed. If so, a stringent selection of case reports is needed to justify the precious space devoted to them in any current issue of a reputed medical journal.

The author and reader must always ask themselves if this report makes a contribution to the first line of evidence in medicine. If not, let us forget it.

A case report certainly represents a very limited contribution to evidence in medicine, but it is also precisely the point at which the cascade of evidence begins. Therefore, we must ensure that our case reports are prepared correctly.

REFERENCES

1. Yackee J, Lipson A, Wasserman AG. Electrocardiographic changes suggestive of cardiac ischemia in a patient with esophageal food impaction. 'A case that's hard to swallow'. *JAMA*,1986;**255**:2065–6.
2. Riesenberg DE. Case reports in the medical literature. *JAMA*,1986;**255**:2067.
3. Swiryn S, Hueter DC. The electrocardiogram in esophageal impaction. *JAMA*,1986;**255**:2067–8.

Assessing the evidence from multiple observations: case series reports and systematic reviews of cases

Assessing the evidence from multiple case observations: case series reports and systematic reviews of cases

Clinical case series are like the military cohorts at the time of the Roman Empire, marching steadily to an uncertain fate. However, it should be easier for us to predict our patient's outcome than the destiny of the warriors under the Centurions' commands.

5.1 WHAT ARE CLINICAL CASE SERIES?

As unusual and rare as some cases may be, they may actually reappear. For some, a case report is limited to one case only[1]. For others, two cases constitute a case report and three[2] to ten[3] cases or more make up a **case series**. Nevertheless, no rule governs the minimum number of cases that form a **set of cases** or a **case series**.

> The study of more than one case has two main objectives:
> - to determine the prevailing characteristics of a given set of patients; and/or
> - to determine the prevailing outcomes for these patients.

For our assessment, let us state that *case series are very different from classical occurrence (descriptive) studies in epidemiology*. In occurrence studies, the number of cases is usually linked to some larger groups of individuals from which they originate. For example, if 10 cases of cancer in a given year are observed in a community of 100 000 subjects, this event may be presented as a **rate** of 10/100 000, where the numerator (10) relates to the denominator of 100 000. In our terms, *case series studies are studies of numerators only*. Denominators in case series studies are most often unknown or extremely difficult to define.

In addition to the 'numerator focus' of case series, only one set of patients is under study. No control group or controlled assignments of patients to clinical manoeuvres is involved. Case series, then, bear little resemblance to observational aetiological research or clinical trials.

It is also hard to ascertain if the assembled cases truly exist or if they are more or less representative of some other cases beyond our grasp.

Nevertheless, aside from these inherent limitations, case studies are often the only source of information about the problem of interest. Therefore, they must be correctly structured, observed, analysed and interpreted within their own limits. Case series are neither proofs of a cause (aetiology) nor proofs of efficacy or effectiveness of treatment (especially in cases when even historical controls are not available).

The presentation of a limited number of cases [4,5] indicates that the first case is probably not unique, and that a better descriptive study might or should be attempted. Such cases can be considered in epidemiological terms as **index cases**, beyond which the study should be expanded. Their presentation must be acceptable from the point of view of **clinimetrics**[6].

For example, Selzer[7] reports several cases of multiple personality. In psychiatry, as in toxicology, clinical microbiology or environmental studies, reports of a limited number of cases are often the only ones available.

Case series are usually presented in one of two forms:
- as review articles[8]; or
- as descriptive studies[9] with all the above-mentioned limitations.

Similarly to any other area of clinical and epidemiological research, case series studies must be prepared according to certain rules. The limitations of case series reports should not be used as an excuse to avoid the rigour of thought required for their presentation. In reality, problems generally arise as a result of the mediocrity of the author's work or the lack of skill displayed by the reader with regards to the reading and understanding of case series reports[10].

5.2 CLASSIFICATION OF CASE SERIES STUDIES

Case series studies may be classified in several ways, as outlined below.

1. **According to the origin of the cases**. The authors may:
   - report cases that they have observed themselves; or
   - assemble case series from several clinical sites; or
   - assemble case series from the literature and present them either in a somewhat haphazard way or systematically as a kind of *'meta-analysis of cases'*.

2. **According to the number of case examinations in time**. The case series can represent:
   - a *cross-sectional study*; i.e. an instant portrait of case characteristics; or
   - a *longitudinal study*, usually based on a single cohort only. Repeated measurements in this situation allow a better understanding of the clinical course of cases and their outcomes.

3. **According to the number of case series involved.** In this situation:
   - either one set of patients is studied in order to obtain their clinical portrait; or
   - two or more case series may be compared in a framework of analytical, observational or quasi-experimental design (different treatments for each series etc.).

A combination of these three axes of classification may be found in each case series report and in the systematic review of cases/case series.

5.3 METHODOLOGICAL ASPECTS OF CASE SERIES REPORTS

For Huth[8], case series analyses are a kind of hybrid paper *'based on a retrospective study of case records, usually cases collected in one institution. The cases may be described in short case reports that are followed by such generalizations as can be drawn from these cases and, perhaps, from similar cases in the literature. In this format the paper has much of the character of a single case report'.* Other papers are written deductively; i.e. to answer some pre-study formulated question, in a form that is closer to more classical observational or experimental aetiological studies.

In general, a case series report is either a sort of a review article that conforms to the corresponding general rules[11] or an occurrence study as mentioned above.

5.3.1 Presenting case series and reviewing the literature

The order of presentation can vary. Either the problem is explained first (for example, pregnancy in Parkinson's disease[12]) and the case report (author's own experience) follows, or the reverse occurs. Although the latter situation is much more prevalent, both sequences are acceptable.

In general, the **presentation of cases** must follow the rules specified in Chapter 3 of this book.

The **review of the literature** is an additional methodological challenge. It should be based preferably on the **systematic review principles and methodology** outlined in other sources[13–15]. The review of the literature must be **evidence-based**, i.e. founded on a uniform and structured evaluation of original evidence from various sources (e.g. a MEDLINE search with terms used in the search), followed by an equally structured synthesis of findings and an assessment of the heterogeneous results obtained from original studies as a basis for clinical decision recommendations.

The review of the literature can cover the pathology under study and its management, or the methodology of the study, depending on the objectives, questions and focus of the case report or case series report. Journal editors should not be too miserly with their space if the literature review is well done.

5.3.2 Establishing a portrait of cases: cross-sectional descriptions

5.3.2.1 Reviewing and summarizing several cases

When regrouping several cases, as in any other systematic review, identical information should be sought and reported from one case to another. The basic information should be presented in the form of an **evidence table**, a technique well known to meta-analysts and to those involved in the systematic review of evidence in health care.

In such tables, the presence or absence of a preselected set of information is *systematically* compared from one clinical case to another whether the cases are from the same source, from multiple sources in the literature, or obtained through clinical case linkages with different hospitals, services etc.

This kind of **checklist of case characteristics** can be included in a cross-sectional study, as in Cole's review of Charles Bonnet hallucinations in psychiatry[16] illustrated in Table 5.1.

The checklist may also be used in longitudinal case series outcome studies, such as Gesundheit *et al.*'s[17] case series report (presentation and follow-up) of nine patients suffering from thyrotropin-secreting pituitary adenoma. Table 5.2 summarizes their findings and approach. This study was also performed according to a pre-established research protocol that is not always present in case series reports.

Table 5.1 – Evidence table of clinical case series in psychiatry: 13 cases of Charles Bonnet hallucinations

		SYSTEMATIC LISTING OF PATIENT CHARACTERISTICS*					OUTCOME*
Age	Sex	Cognitive Impairment	Visual Impairment	Living Alone	Content of Hallucinations	Insight	Cognitive State after One Year
82	Female	Minimal	Severe	Yes	Two cats, five kittens	No	Unknown
73	Female	Mild dementia	Mild	No	Strangers (children)	No	Unknown
78	Female	Mild dementia	None	No	Landlord and family	No	Unknown
77	Female	Minimal	Mild	Yes	Strangers (young men)	No	Severe, rapid decline
82	Female	Minimal	Severe	Yes	Strangers (children and adults)	Yes	Mild decline
71	Female	Mild dementia	None	Yes	Strangers (adults)	No	Severe, rapid decline
84	Male	Minimal	Moderate	Yes	Strangers (adults)	Partial	Unchanged
84	Female	None	Mild	Yes	Strangers (children and adults)	No	Severe, rapid decline
71	Male	None	Mild	No	Strangers (men)	No	Severe, rapid decline
75	Female	Mild dementia	Moderate	Yes	Strangers (adults)	No	Unknown
72	Female	None	Mild	No	Animals, human faces	Yes	Unchanged
80	Female	Mild dementia	None	Yes	Strangers (adults)	No	Unknown
70	Female	None	None	Yes	One cat	Yes	Unchanged

Source: Ref. 16 – Reproduced with permission.
*didactic additions

Table 5.2 – Outline of an evidence table of a case series in medicine
Abridged from a full review of nine cases of TSH-secreting pituitary adenomas

Case No	Patient Age*	Sex	Initial diagnosis	Initial therapies	Date of diagnosis of TSH adenoma	Associated endocrine abnormalities	Subsequent therapies and operative findings	Result after therapies
1	51	M	Hypogonadism (1969) Graves disease (1978)	Androgens thionamides	1982	Increased LH increased FSH	Pituitary surgery: invasive macroadenoma; radiation	Persistent tumor, monocular blindness
⋮								
3	43	F	Nonfunctioning pituitary tumor (1956)	Transfrontal hypophysectomy (1956) Radiation (1956)	1983	Increased growth hormone, clinical acromegaly	Pituitary surgery: invasive macroadenoma	Death
⋮								
5	35	F	Graves disease (1976)	Thionamides	1983	Increased prolactin	Pituitary surgery: invasive macroadenoma	Persistent tumor
⋮								
9	40	M	Graves disease (1980)	Thionamides, total thyroidectomy	1986	Increased prolactin	Pituitary surgery: microadenoma	Biochemical normalization
⋮								

Source: Ref. 17 – Reproduced with permission
*Age at time of diagnosis of TSH adenoma. TSH = thyroid-stimulating hormone; LH = luteinizing hormone; FSH = follicle-stimulating hormone.

In contrast to single case reports that are mainly inductive, case series reports are a form of deductive research. Questions and hypotheses are formulated beforehand and the case series review is carried out to answer specific questions and to test hypotheses.

5.3.2.2 Case series with or without denominators

In toxicology, surveillance of unique cases is often used to better understand the toxic effects of various substances. For example, the increasing and often unjustifiable popularity of unconventional therapies has also produced new risks in the form of seemingly 'natural' and hence 'innocuous' treatments that are supposed to do only good. Anderson *et al.*[18] studied the toxicity of pennyroyal, a herbal abortifacient with potentially lethal hepatotoxic effects, available in some health food stores.

Four cases observed longitudinally by other authors and eighteen previous case reports provided information about the toxicity of this herbal medicine, known since Roman times. The objective of this case series study was to better understand the toxicity of the substance rather than to study the occurrence of exposure and intoxication in a well-defined target population as denominator.

In other situations, denominators are available. Braun and Ellenberg[19] provide a descriptive epidemiology of adverse effects of immunizations per population of 100 000 in the United States. Their case series involved 38 787 adverse effect cases.

5.3.2.3 Case series from a single source

Case series originating from a single source such as one hospital, department, laboratory etc. have two obvious advantages: first, the experience with multiple cases usually elicits more direct questions and second, hypotheses and their study may be carried out deductively. Gesundheit *et al.*'s[17] mainly qualitative study of clinical and paraclinical features of thyrotropin-secreting pituitary adenomas falls into this category.

Other studies, however, highlight a particular team's experience, such as D'Abrigeon et al.'s[20] observation of five cases of papillomatosis of the biliary tract (papillary adenoma or papillary adenocarcinoma).

A predetermined protocol is highly desirable. For example, Braun et al.'s[21] study of snowmobiling-related brachial plexus injuries was based on a systematic retrieval of sociodemographic and behavioural characteristics of injured individuals, injury accident characteristics, diagnostic methods, associated injuries, signs and symptoms of brachial plexus involvement, and follow-up data, when available.

The enormous advantage offered by case series from a single source is the possibility of selecting and describing cases according to uniform clinimetric criteria of case detection, diagnosis, selection, measurement of selected variables, and final categorization and interpretation.

A systematic description of cases should lead to their synthesis as a basis for new hypotheses and more justified indications for the practice concerned. Dennis et al.[22] found a higher than expected number of elderly patients suffering from systemic lupus erythematosus with associated psychiatric problems such as confusional states, dementia and depression. A more common occurrence of this problem in the elderly as well as diagnostic criteria and strategies of detection were hypothesized.

Singh and Scheld[5] used their cases of rhabdomyolysis to research the infectious compared with the non-infectious aetiology of this clinical entity.

5.3.2.4 Case series from the literature

Sometimes, casuists are limited in their studies to case series found in the literature. In these situations, the authors may identify their own and rare observations first, for example a malignant intracerebral nerve sheath tumour[23], a pulmonary blastoma[24] or a cytomegalovirus pericarditis[25]. Such first personal observations form the index case, to which records and reports from the literature can be linked.

Case series constituted in such a way have a major disadvantage with regards to the variable heterogeneity of definitions, measurement criteria, diagnostic and therapeutic strategies and decisions, and follow-ups. It is the authors' responsibility to indicate in their studies the value and limitations of their findings stemming from this type of problem inherent in case series assembled 'from the outside'.

5.3.3 Follow-up of outcome studies or longitudinal case series

5.3.3.1 Follow-up of observational case series

When there is more than one reported examination of patients, longitudinal studies are produced. In contrast, instant or 'snapshot' pictures of cases are based on one measurement and evaluation only.

Longitudinal studies, as repeated examinations of case series in time, focus on various disease outcomes. Survival, longevity, cure, remission, pain resolution, normal organ function, return to normal values of biological variables, resumption of everyday life activities, and social and professional reintegration of the patient are just some of the challenges of outcome research. Their particular application to dentistry by Anderson[26] is worthy of attention.

Broadly defined, an **outcome** can be *'any possible result that may stem from exposure to a causal factor, or from preventive and therapeutic interventions: all identified changes in health status arising as a consequence of the handling of a health problem'* [27].

Currently, many case reports and case series reports lack an *a priori* definition and a choice of outcome. Justification of a chosen outcome is rare, especially with regards to clinical decisions that have to be modified or made.

However, the number of outcome-focused case series is increasing. For example, Tiffany et al.[28] followed a series of patients with refractory arrest rhythms. In contrast to the contraindication of thrombolytic therapy in patients receiving cardiopulmonary resuscitation, the authors hypothesized that this intervention in selected patients might facilitate their outcome in terms of the spontaneous return of circulation.

Bloom et al.[29] carried out a prospective study of visual acuity in 147 AIDS patients with cytomegalovirus retinitis. These authors concluded that standard treatment for this condition minimizes loss of vision and might protect previously uninfected eyes.

Hence, outcome case series studies of one group of patients are similar in their basic organization (and they should be) to Phase I or early Phase II clinical trials that aim to establish how healthy (Phase I) or diseased (Phase II) individuals will respond to treatment[30]. (NB To complete the picture, let us remember what follows Phases I and II. Phase III is a randomized controlled clinical trial based on the comparison of two or more groups. Phase IV consists of trials without patient preselection, while Phase V is based on trials that include patients with various co-morbidities and co-treatments for the latter.)

> If Phase I or II clinical trials require an adequate protocol, so too do longitudinal case series studies focusing on various outcomes. From the methodological point of view, the latter must be considered as early phases of clinical modalities evaluations by experimental means.

5.3.3.2 Follow-up of experimental case series

In an experimental design, the researcher must decide which patient will be treated and which treatment will be offered (a certain drug, placebo, nothing etc.).

Case series studies can also be organized from an experimental point of view. However, only alternative designs to classical randomized controlled clinical trials are possible for obvious reasons.

An A-B-A design was used by Papageorgiou and Wells[31] to study the effects of attention training on the improvement of affect, illness-related behaviour and cognition in hypochondriasis sufferers. Patients acted as their own control. 'A' was the treatment under study, while 'B' was the withdrawal of the treatment period. The results obtained suggested that hypochondriacs might also benefit from the therapy under study.

Inherent characteristics of case series studies, such as the reduced number of patients and their heterogeneous origins and the management of such studies, limit the conclusions that can be drawn from these studies. Nevertheless, although case series studies are often the only feasible option, this should not mean that case series studies can be incorrectly prepared and interpreted.

5.3.4 'Meta-analysis' or systematic reviews of cases and case series

In the past twenty years, an increasing number of researchers in psychology, education and the health sciences have felt the need to integrate independent findings from different sources and studies focusing on the same topic or research problem. In the field of quantitative research, an 'analysis of analyses' or **meta-analysis** has been defined, methodologically developed and used increasingly, especially in the area of experimental research and clinical trials.

In the health sciences, the first textbooks on meta-analysis[32,33] appeared about ten years ago. Since then, a rapid development of quantitative methodology in meta-analysis and a widening of its applications have positioned this type of systematic review of evidence well beyond the scope of these first works. However, occasional updates[34] are worthy of attention.

We have defined **meta-analysis in medicine and allied health sciences** as *'a systematic, organized and structured evaluation and synthesis of a problem of interest, based on the results of many independent studies of that problem (disease cause, treatment effect, diagnostic method, prognosis etc.). Results of different studies become a new unit of observation and the subject of study is a new cluster of data, similar to groups of subjects in original studies. It is a 'study of studies' or 'epidemiology of their results'. Its objectives are: to confirm information, to find errors, to search for additional findings (induction) and to find ideas for further research (deduction)'* [13,32,33]. Meta-analysis must have both qualitative and quantitative components[13,32,34].

By substituting individual case studies or case series reports as a unit of observation, is there a need for the **meta-analysis of cases and of case series reports**?

Just as the results of original studies are heterogeneous, so are those of clinical case reports and case series reports. Is there a way to integrate them to obtain the most accurate picture of the disease and its control, especially when only isolated case reports are available?

The idea is relatively new, but it is a logical evolution of thinking in qualitative and quantitative research. Simultaneously but independently, conceptual proposals have appeared in qualitative research[35–37] as well as in the first *'meta-analyses of case reports and case series reports'* [38–41].

Jensen and Allen[35] investigated wellness and illness based on 112 qualitative studies of individuals suffering from a wide variety of health problems. They called their synthesis and inductive interpretation 'meta-analysis'. Criteria for the selection of qualitative studies for aggregation were proposed at the same time by Eastbrooks, Fields and Morse[36]. 'Interpretive meta-synthesis' was used later by the same Jensen and Allen[37] to describe this kind of endeavour. Independent of the terms selected, an attempt was made to outline a better method of preparing aggregate case reports.

In clinical oncology, Nayfield and Gorin[42] retrieved from the literature twenty-one cases of tamoxifen-related ocular toxicity in breast cancer sufferers: one cohort and five small cross-sectional studies of occurrence of ocular findings in tamoxifen recipients. Their review of these cases produced a clearer picture of the nature and distribution of the toxicity, as well as of the severity of the ocular findings, which is useful in the management of tamoxifen-treated patients. They recognized the difficulty of attributing ocular findings to tamoxifen and other competing causes of retinal, macular and corneal abnormalities.

Fraser, Grimes and Schultz[38] evaluated an 'overload' of antigen in expectant mothers which mimicked the presentation of fetal antigen during pregnancy as a therapy for recurrent spontaneous abortion. Two meta-analyses, one aggregation of case series findings and one of three clinical trials, both provided evidence against its use until future randomized controlled clinical trials proved otherwise.

Nordin[39] made a 20-year literature review of both case reports and case series reports of primary carcinoma of the fallopian tube. This author was able not only to outline more appropriate patient characteristics, but also to hypothesize on the underdiagnosing of cases, causes of treatment failure, lack of controlled trials and usefulness of a 'second look laparotomy' for monitoring disease response to the treatment given (extensive debulking surgery and adjuvant platinum-based combination chemotherapy).

Drenth et al.[40] retrieved from the literature 126 articles on erythermalgia. Nine children in the articles met the inclusion criteria for the meta-analysis of cases. Descriptive characteristics of patients, spectra of clinical management, and treatment results were quantified across this set of observations. It was found that erythromelalgia might be associated with elevated blood pressure. Prognosis and treatment effect across the cases seemed different in primary and secondary erythermalgia. However, conclusions were limited due to the nature of a single case group study retrieved from the literature.

Cook et al.[41] also reviewed both case series and case reports in studying the outcomes of traumatic optic neuropathy in patients classified according to the standardized grading system. Recovery was related to the severity of the initial injury.

Obviously, none of these meta-analyses of cases is explanatory. Only prevalent case characteristics and case outcomes can be better assessed at this level of clinical research.

Establishing prevalent characteristics or average or typical values of observations in cases is similar to establishing **typical odds ratios** or other summary characteristics in clinical trials or in aetiological observational research. Such a **quantitative meta-analysis** usually follows a **qualitative meta-analysis,** i.e. *'a method of assessment of the importance and relevance of medical information coming from several independent sources through (by) a general, systematic and uniform application of pre-established criteria of acceptability of original studies representing the body of knowledge of a given health problem or question'* [13,32,34].

This kind of assessment of original studies in meta-analysis also applies to the systematic review of cases. However, we should explore and develop such **qualitative meta-analysis of cases** before regrouping them by other means. Until now, this has not been attempted. No one has explored the quality or the completeness of information from one case to another even though this would enable us to better understand if such a qualitative review of cases is possible and realistic or not. Should this be a requirement in the future? The answer is probably.

By comparing case integration to a remarkable set of meta-analyses of clinical trials, it can be seen that research synthesis of case reports and case series reports marks only the beginning. The future will show how far we can go in this area.

5.4 CONCLUSIONS

Anyone who searches for evidence of what is helpful or harmful to patients is not necessarily satisfied with conclusions drawn from case reports and case series reports[43,44]. But if no other more adequate evidence is available, the best of this first line of evidence must be taken into account.

If this line of evidence is limited, and it often is, this is not a justification to reject it. On the contrary, it is a justification to use the evidence as well as possible within its own limits in a generally descriptive observational study.

Califf[45] states: '. . .*Case series without a control group will remain interesting because of the intrinsic importance of observation in medicine. Although individual case reports should never be taken as definitive evidence that practice should be changed, the importance of astute, appropriate bedside observation cannot be overestimated'*.

If an 'astute clinical observation' is at the root of case reports, should we teach young clinicians 'astuteness' and, if so, how should this be done?

Let us also realize that clinical case reporting is the clinician's most frequent exposure to medical research. The vast majority of practitioners will never carry out cohort or case control studies or complex clinical trials. Nevertheless, they should be as proficient clinical case reporters as possible.

The hierarchy of evidence proposed and adopted in North America[45–47] places descriptive studies at a very low level:

The strongest:	I	Randomized clinical controlled trials (at least one).
	II - 1	Well-designed non-randomized controlled trials.
	II - 2	Well-designed cohort or case control studies (preferably multicenter).
	II - 3	Multiple time series or place comparisons with or without interventions (including uncontrolled experiments, like at the onset of the antibiotic era).
The weakest:	III	Opinions of respected authorities, based on clinical experience, **descriptive studies** or reports of expert committees.

Sackett's[43] levels of evidence show a similar structure:

The strongest:	Level I	Randomized trials carrying low alpha and beta errors.
	Level II	Randomized trials carrying high alpha and beta errors.
	Level III	Non-randomized concurrent cohort comparisons (simultaneous observational comparisons of more than one group, e.g. treated and not treated subjects).

| | Level IV | Non-randomized historical co-hort comparisons of more than one group, with control groups from the past originating either from the same institution or from the literature. |
| **The weakest:** | Level V | **Case series without controls.** |

Case reports and case series reports may be the 'lowest' or the 'weakest' level of evidence, but they often remain the 'first line of evidence'. This is where everything begins.

Let us note that such hierarchical classifications are not unequivocally accepted. Is a poorly designed randomized controlled clinical trial really better than an impeccable case control study? Does evidence from studies giving information about an 'average patient' apply to other patients, belonging to various diagnostic or prognostic subgroups? What should be done concerning decisions focusing on soft data or outcome measures that have not yet been evaluated by a soft data focused clinical trial (patient comfort, well-being etc.)[48]? What should be done if potentially adverse effects and undesirable outcomes from co-morbidities and their treatment are one of the main concerns?

All levels of evidence should be closely examined. How any level and hierarchy of evidence should be applied is the responsibility of its users.

Nevertheless, we can agree that in case series without controls *'the reader is simply informed about the fate of a group of patients. Such series may contain extremely useful information about clinical course and prognosis but can only hint at efficacy'* [43]. Obviously, the 'information' and the 'hint' should be as accurate as possible.

It is for this particular reason that case experience, case reports and case series reports must be as correctly carried out and presented as any other line of evidence. The *quality of data* and the *scope of interpretations and recommendations* are perhaps their most important assets. *The eligibility of record linkage of cases from independent sources* should also be respected. If cases originate from a one-source observation, they must be as sound methodologically as Phase II clinical trials.

If case reports and case series reports are of high quality, they really shouldn't be constrained by space limitations in medical journals. They do not deserve this fate. If space is excessively limited, the reader's only recourse is usually to hypothesize about the quality and completeness of the case report itself and its message. The 'underdogs of medical evidence' should be raised at least to the level of 'foot soldiers of evidence' who shoot as well as their guns let them.

Let us conclude on a lighter note. Was our historical example of the Immaculate Conception really a unique case report? Others[49] have assembled an impressive array of similar cases from diverse historical and religious contexts: Adonis, Zoroaster, Krishna, Mithra, Gautama Buddha, Dionysus, Quirmus, Attis and Indra. In fact, an evidence table could help identify known virgin mothers, including Maya (Gautama Buddha), Nama (Attis) and Devaki (Krishna), and the geographical location of such cases (Babylon, India, Tibet, Phrygia, Greece, Rome) would favour the occurrence of the event in question in the temperate climates of Asia or along the shores of the Mediterranean. Two additional births, those of Dionysus and Mithra, even took place in stables. (Incidentally, the latter was also born on December 25th.) Such an apparently more frequent occurrence than expected no doubt requires a more complete case series report, sources of data permitting, especially since the location of events in time suggests a sudden drop in the number of new cases in the past few centuries!

Our case series reports should be better and more complete.

REFERENCES

1. Coates M-M. Writing for publication: Case reports. *J Hum Lact*,1992;**8**(1):23–6.
2. Instructions for authors. *Obstet Gynecol*,1994; **83**:six title pages (unnumbered).
3. Simpson RJ Jr, Griggs TR. Case reports and medical progress. *Persp Biol Med*,1985;**28**:42–6.
4. Kamarulzaman A, Briggs RJS, Fabinyi G, Richards MJ. Skull osteomyelitis due to Salmonella species: Two case reports and review. *Clin Infect Dis*,1996;**22**:638–41.
5. Singh U, Scheld WM. Infectious etiologies of rhabdomyolysis: Three case reports and review. *Clin Infect Dis*,1996;**22**:642–9.
6. Jenicek M. Making clinical pictures or describing disease in an individual, pp. 125–136 in: *Epidemiology. The Logic of Modern Medicine.* Montreal: Epimed International, 1995.
7. Seltzer A. Multiple personality: A psychiatric misadventure. *Can J Psychiatry*, 1994; **39**:442–5.
8. Huth EJ. The review and the case-series analysis, pp. 64–68 in: *How to Write and Publish Papers in the Medical Sciences.* Philadelphia: iSi Press, 1982.
9. Jenicek M. Picturing disease as an entity. Describing disease occurrence in the community, pp. 136–145 in: *Epidemiology. The Logic of Modern Medicine.* Montreal: Epimed International, 1995.
10. Grimes DA. Technology follies. The uncritical acceptance of medical innovation. *JAMA*,1993:**269**:3030–3.
11. Milne R, Chambers L. Assessing the scientific quality of review articles. *J Epidemiol Comm Med*,1993;**47**:169–70.
12. Hagell P, Odin P, Vinge E. Pregnancy in Parkinson's disease: A review of the literature and a case report. *Movement Disorders*,1998;**13**:34–8.
13. Jenicek M. Meta-analysis in medicine: Where we are and where we want to go. *J Clin Epidemiol*,1989;**42**:35–44.
14. Mulrow C, Oxman A. *How to Conduct a Cochrane Systematic Review.* San Antonio: San Antonio Cochrane Center, 1996.
15. The Potsdam International Consultation on Meta-Analysis. Potsdam, Germany, March 1994. Edited by WO Spitzer. *J Clin Epidemiol* (Special Issue), 1995;**48**:1–171.
16. Cole MG. Charles Bonnet hallucinations: A case series. *Can J Psychiatry*,1992;**37**:267–70.
17. Gesundheit N, Petrick P, Nissim M, Dahlberg PA, Doppman JL, Emerson CH, Braverman LE, Oldfield EH, Weintraub BD. Thyrotropin-secreting pituitary

adenomas: Clinical and biochemical heterogeneity. Case reports and follow-up of nine patients. *Ann Intern Med*,1989;**111**:827–35.

18. Anderson IB, Mullen WH, Meeker JE, Khojasteh-Bakht SC, Oishi S, Nelson SD, Blane PD. Pennyroyal toxicity: Measurement of toxic metabolite levels in two cases and review of the literature. *Ann Intern Med*, 1996;**124**:726–43.

19. Braun MM, Ellenberg SS. Descriptive epidemiology of adverse events after immunization: Reports to the Vaccine Adverse Event Reporting System (VAERS), 1991–1994. *J Pediatr*,1997;**131**:529–35.

20. D'Abrigeon G, Blanc P, Bauret P, Diaz D, Durand L, Michel J, Larrey D. Diagnostic and therapeutic aspects of endoscopic retrograde cholangiography in papillomatosis of the bile ducts: analysis of five cases. *Gastrointest Endosc*,1997;**46**:237–43.

21. Braun BL, Meyers B, Dulebohn SC, Eyer SD. Severe brachial plexus injury as a result of snowmobiling: A case series. *J Trauma Injury Infect Crit Care*,1998;**44**:726–30.

22. Dennis MS, Byrne EJ, Hopkinson N, Bendall P. Neuropsychiatric systematic lupus erythermatosus in elderly people: a case series. *J Neurol Neurosurg Psychiatry*, 1992;**55**:1157–61.

23. Sharma S, Abbott RI, Zagzag D. Malignant intracerebral nerve sheath tumor. A case report and review of the literature. *Cancer*,1998;**82**:545–52.

24. Cutler GS, Michel RP, Yassa M, Langleben A. Pulmonary blastoma. Case report of a patient with a 7-year remission and review of chemotherapy experience in the world literature. *Cancer*,1998;**82**:462–7.

25. Campbell PT, Li JS, Wall TC, O'Connor CM, Van Trigt P, Kenney RT, Melhus O, Corey GR. Cytomegalovirus pericarditis: A case series and review of the literature. *Am J Med Sci*,1995;**309**(4):229–43.

26. Anderson JD. The need for criteria reporting treatment outcomes. *J Prosthet Dent* 1998;**79**:49–55.

27. *A Dictionary of Epidemiology*, 3rd Edition. Edited by JM Last. New York: Oxford University Press, 1995.

28. Tiffany PA, Schultz M, Stueven H. Bolus thrombolytic infusions during CPR for patients with refractory arrest rhythms: Outcome of a case series. *Ann Emerg Med* 1998;**31**:124–6.

29. Bloom PA, Sandy CJ, Migdal CS, Stanbury R, Graham EM. Visual prognosis of AIDS patients with cytomegalovirus retinitis. *Eye*,1996;**9**:697–702.

30. Jenicek M. Phases of evaluation of treatment, pp. 213–215 in: *Epidemiology. The Logic of Modern Medicine*. Montreal: Epimed International, 1995.

31. Papageorgiou C, Wells A. Effects of attention training on hypochondriasis: a brief case series. *Psychol Med*,1998;**28**:193–200.

32. Jenicek M. *Méta-Analyse en Médecine. Evaluation et Synthèse de l'Information Clinique et Épidémiologique. (Meta-analysis in Medicine. Evaluation and Synthesis of Clinical and Epidemiological Information.)* St Hyacinthe and Paris: Edisem and Maloine, 1987.

33. Petitti DB. *Meta-Analysis, Decision Analysis, and Cost-Effectiveness Analysis. Methods of Quantitative Synthesis in Medicine.* Monographs in Epidemiology

and Biostatistics, Volume 24. New York and Oxford: Oxford University Press, 1994.

34. Jenicek M. Meta-analysis in medicine. Putting experiences together, pp. 267–295 in: *Epidemiology. The Logic of Modern Medicine*. Montreal: Epimed International, 1995.

35. Jensen LA, Allen MN. A synthesis of qualitative research on wellness–illness. *Qualit Health Res,*1994;**4**:349–69.

36. Estabrooks CA, Filed PA, Morse JM. Aggregating qualitative findings: An approach to theory development. *Qualit Health Res,*1994;**4**:503–11.

37. Jensen LA, Allen MN. Meta-synthesis of qualitative findings. *Qualit Health Res,*1996; **6**:553–60.

38. Fraser EJ, Grimes DA, Schultz KF. Immunization as therapy for recurrent spontaneous abortion: A review and meta-analysis. *Obstet Gynecol,* 1993;**82**:854–9.

39. Nordin AJ. Primary carcinoma of the fallopian tube: A 20-year literature review. *Obstet Gynecol Survey,*1994;**49**:349–61.

40. Drenth JPH, Michiles JJ, Özsoylu S. Erythermalgia Multidisciplinary Study Group. Acute secondary erythermalgia and hypertension in children. *Eur J Pediatr,*1995;**154**:882–5.

41. Cook MW, Levin LA, Joseph MP, Pinczover EF. Traumatic optic neuropathy. A meta-analysis. *Arch Otolaryngol Head Neck Surg,*1996;**122**:389–92.

42. Nayfield SG, Gorin MB. Tamoxifen-associated eye disease: A review. *J Clin Oncol,* 1996;**14**:1018–26.

43. Sackett DL. Rules of evidence and clinical recommendations. *Can J Cardiol,*1993;**9**: 487–9.

44. Peipert JF, Gifford DS, Boardman LA. Research design and methods of quantitative synthesis of medical evidence. *Obstet Gynecol,*1997;**90**:473–8.

45. Califf RM. How should clinicians intepret clinical trials? *Cardiol Clinics,* 1995;**13**:459–68.

46. Canadian Task Force on the Periodic Health Examination. *The Canadian Guide to Clinical Preventive Health Care*. Ottawa: Health Canada, 1994.

47. Preventive Services Task Force. *Guide to Clinical Preventive Services*, 2nd Edition. Baltimore: Williams & Wilkins, 1996.

48. Feinstein AR, Horwitz RI. Problems in the 'evidence' in 'evidence-based medicine'. *Am J Med,*1997;**103**:529–35.

49. Knight C, Lomas R. *The Hiram Key*. London: Arrow Books Ltd, 1997.

CHAPTER 6

What next?

CHAPTER 6

What next?

Foretelling the future of clinical case reports is like making their prognosis. This prognosis is now more promising than ever, given the advances in casuistics made possible by clinimetrics and epidemiological thinking in clinical medicine and by the awareness that state of the art clinical case reports open Pandora's box and Croesus' trove of further medical research and knowledge

In the past ten years, the field of medical casuistics has been significantly refined and restructured. The spontaneity and compulsiveness of many casuists who used to provide information simply because they felt that they should, have been abandoned in favour of a more rigorous approach.

6.1 HOW CLINICAL CASE REPORTS SHOULD BE EXAMINED TODAY

The most important new paradigm of clinical case reports is the view that they open the gate to further more refined and more complete research. Clinical case reports are also a part of an increasing equilibrium between qualitative and quantitative research. Predominantly quantitative research usually follows case reports.

Most of us, to a certain degree, have been trained in epidemiology and have some knowledge of it as the basis for quantitative research and logic in medicine. Nevertheless, we may feel uncertain about recent methodological developments in qualitative research, especially from areas other than the health sciences. We also experienced those same feelings when meta-analysis was introduced, and when even the basic paradigm of medicine shifted from a deterministic view to a probabilistic view of the health phenomena that surround us. This is all now fading away.

Clinical case reports are competing for the attention of readers and for space in medical journals. The reports must now be as good as any other type of medical information and research.

The reader's occasional rapid loss of interest in reading a clinical case report can be attributed to several problems:
- Editors of medical journals who do not give authors enough information concerning what is expected of them and their work.

- Authors who do not have the knowledge required to determine the relevance and originality of their case(s), and to present their case(s) properly.
- Readers of clinical case reports who do not know how to draw relevant information from the case reports and how to judge the information's usefulness for their practice and research.

In other words, authors must know why the case should be reported and how to report it, editors must be explicit with regards to the publishing requirements of the case, and readers should be trained to properly interpret the message of a case report.

If a case report or a case series report is not clear enough in terms of:
- the reason for its presentation,
- its clinimetric explicitness,
- its explanation of facts and actions,
- its recommendations for practice and research stemming from the case(s) experience,

then the reader will remain perplexed, blasé and disinterested.

Clinical case reports also offer an invaluable additional asset. If they are prepared in a structured and organized manner, their authors learn how to organize and structure their own thoughts and practice. Hence, the clinical report is an important teaching tool[1-3], both for the author and the reader.

In public health and field epidemiology, case reports based on index cases 'of something bigger' remain powerful generators of hypotheses in the search for causes of outbreaks, epidemicity or pandemicity of disease, or in considering the need to evaluate the efficacy and effectiveness of health programmes.

Theoretically speaking, the outbreak of an epidemic is a 'situation', i.e. a 'case', with respect to the qualitative research outlined in Chapters 2 and 3. Handling and reporting a disease outbreak is therefore a 'case study' with additional important quantitative components. Its methodological rules go well beyond the scope of this reading and have been discussed in other sources[4].

Clinical research usually requires that a study deals with a considerable number of patients in order for it to be 'statistically powerful' and to minimize beta errors. Does this mean that case reports are 'powerless'? The answer is yes and no. In reality, they are undeniably strong since they lead to further observational, aetiological and experimental research. One situation should not exclude the other[5,6].

Cases are 'uncontrolled experiments'. They do not provide proofs, but they generate hypotheses. It is impossible to underestimate them for this latter reason. It should also not be forgotten that Darwin's and Freud's theories were founded on case observations, studies and reports and not on what we would today call 'serious aetiological and experimental research'[6].

Maple syrup disease was explained on the basis of first case observations and reports of progressive neurological degeneration and peculiarly odoriferous urine. These reports led to the hypothesis of a genetic error of metabolism, which was later confirmed[7].

Early reporting of important life- and health-threatening clinical cases also allows rapid emergency control measures to be put in place. Although aetiological confirmations follow later, at times some decisions simply can't wait. By way of example, one only has to think of the congenital malformations that occurred after thalidomide use[8].

Moreover, cases can lead to a future aggregation with other cases in either observational or quasi-experimental research. They should therefore be well defined in order to serve this purpose. Clinical microbiologists, toxicologists and psychiatrists often depend on them.

6.2 SUGGESTIONS TO ASPIRING CASUISTS

Accumulated from one generation to the next, the wealth of experience contained in clinical case reporting will continue to increase with time. To improve these reports even further, recently borrowed concepts from the areas of epidemiological thinking and methodological refinements in clinimetrics must be mastered and put to good use. These concepts are reflected in the following 'Ten Commandments':

- Persevere.
- Look for something new and relevant, worthy of sharing.
- Learn, adopt and use clinimetrics[9–11]. Observe and describe thoroughly. Correctly record what you have heard, seen, touched, smelt and read.
- Reason in terms of epidemiology.
- Temper your interpretations.
- Avoid exaggerated conclusions.
- Consider recommendations that stem from and are proportional to the experience of the case study.
- Retain only what is relevant for medicine and beneficial for the patient.
- Consider case series studies as rigorous descriptive studies of diagnosis, follow-up or prognosis without *a priori* statistical considerations. However, work up their design and protocol

according to the rules and standards of Phase II clinical trials[12–16].

- While remaining critical, have fun, and be proud of your accomplishments. You are doing first rate clinical research within the limitations of clinical case reports.

Good clinical case reports are challenging. Often, there are no *a priori* hypotheses of disease aetiology. In addition, there is usually no *a priori* knowledge of the disease itself or of its course and control.

When several cases are regrouped, some heterogeneity is unavoidable. These cases are often detected haphazardly and are observed for different reasons, in different ways. Then, they are subjected to a 'boot camp' of case series research, where the final result expected is 'a nice line of uniformed soldiers marching in time and in line' toward a common purpose and goal. The end-point of such a march should not signify the scientific death of the cohort we assembled in our hearts and minds.

6.3 WHAT REMAINS TO BE DONE

Most of the challenges of medical casuistics are related to its direction and methodology:
- Is it possible to define criteria more completely according to which case should be reported?
- Should we give a different form and structure to case reports, focusing on different problems such as challenging differential diagnosis, clinical management of an unexpected course and outcome or choosing between competing therapeutic and disease management plans?

- Should we make clinicians' and general practitioners' training in medical casuistics and proper clinical case reporting more formal and structured?
- If a clinical case report triggers expected further clinical research, is it logical and necessary to compare results from initial case observations and their interpretation with what is obtained at other stages in the research of the same problem?
- Would it be useful to refine and restructure specific presentation rules for series of cases?

The future will most likely provide answers to most of these questions. In the meantime, if this reading helps make health professionals more aware of the advances and challenges of today's medical casuistics, its objectives will have been met.

6.4   LET US CONCLUDE

Clinical case studies and reports are an inherent and indispensable part of medical research, progress and experience. Moreover, they are also very helpful in maintaining the human aspect of medicine and medical research. Despite knowing all about the numerical facts, probabilities, risks and chances relevant to them, patients will continue to ask their physicians: 'and what about me?'. A 50 per cent case fatality rate may be a fact, but the patient is either dead or alive. In other words, a case remains a case, and a patient a patient, whether in practice or research.

Medical casuists must handle each case as a unique experience and as a broader part of medical knowledge, experience and generalized wisdom. The challenge is to manage both parts of the equation equally well.

As long as medicine exists, physicians will monitor John Doe or Jane Smith's pulse rate and blood pressure. They will also know the pulse rate and blood pressure levels in John and Jane's community. John and Jane are not at all interested in what is happening around them, but they want their blood pressure to be under control in order to avoid any life-threatening complications. Physicians will and must always make distinctions between individual cases and observations and generalizations arising from the study of many cases and individuals[17]. As they focus their decisions on individual cases, they must also aim to control the disease in the community. They must therefore use community experience to treat individual patients. Both approaches, patient- and community-centred, are complementary and necessary.

Is medical casuistics a science? Reiser[18] states that *'the classification of facts, and the recognition of their sequence and relative significance is the function of science'* as well as *'the habit of forming judgement upon these facts unbiased by personal feeling'*. In this light, and this is where the challenge lies, medical casuistics should be a part of the science of medicine.

Is the clinical case report only the first line of evidence in the formal aetiologic research in medicine? It may be more than that.

In 1975, Carol Buck[19] emphasized how hypothetico-deductive reasoning is important for epidemiologists. Thus she confirms Karl Popper's and Thomas Kuhn's philosophy of research. These philosophers of science consider a contradictory finding to be an important 'falsifier' or an overthrow of a current paradigm (*Gestalt switch*).

For Velanovich[20], 'the case report is relatively strong evidence that an existing paradigm is wrong if it provides evidence that does not corroborate the paradigm. However, it is relatively weak evidence for a new hypothesis even though it may be highly corroborative, because of fallacies in judgment associated with the law of small numbers, representativeness, and availability. In some very special cases, the information obtained from a case report may be the impetus for a "Gestalt switch" in a paradigm of medical and surgical thinking'.

Clinical case research remains an important element in maintaining an equilibrium between research based on solid numerical data and research based on individual clinical experience. This is a case of and for 'medical research and practice with a human face'.

Obviously, appropriate measurements, quantifications and categorizations help in individual cases here, as much as in cases beyond the field of medicine, such as human or animal 'beauty' contests. The rewards are either a crown or a clean bill of health.

If the reader still doubts that medical casuistics is a serious endeavour, he should raise his hand and be counted. After that, he should explain his reasoning!

REFERENCES

1. Loschen EL. The resident conference: A method to enhance academic intensity. *J Med Educ*,1980;**55**:209–10.
2. Petrusa ER, Weiss GB. Writing case reports. An educationally valuable experience for house officers. *J Med Educ*,1982;**57**:415–7.
3. Winter R. Edit a staff round. *Br Med J*,1991;**303**:1258–9.
4. Jenicek M. Investigations of disease outbreaks, pp. 189–192 in: *Epidemiology. The Logic of Modern Medicine*. Montreal: Epimed International, 1995.
5. Feinstein AR. An additional basic science for clinical medicine: I. The constraining fundamental paradigms. *Ann Intern Med*,1983;**99**:393–7.
6. Herman J. Experiment and observation. *Lancet*,1994;**344**:1209–11.
7. Simpson RJ Jr, Griggs TR. Case reports and medical progress. *Persp Biol Med*,1985; **28**:402–6.
8. Levine M, Walter S, Lee H, Haines T, Holbrook A, Moyer V for the Evidence-Based Medicine Working Group. Users' guides to the medical literature. IV. How to use an article about harm. *JAMA*,1994;**271**:1615–9.
9. Feinstein AR. *Clinimetrics*. New Haven and London: Yale University Press, 1987.
10. Feinstein AR. An additional basic science for clinical medicine. IV. The development of clinimetrics. *Ann Intern Med*,1983;**99**:843–8.
11. Jenicek M. Identifying cases of disease. Clinimetrics and diagnosis, pp. 79–118 in: *Epidemiology. The Logic of Modern Medicine*. Montreal: Epimed International, 1995.
12. Jenicek M. Phases of evaluation of treatment, pp. 213–215 in: *Epidemiology. The Logic of Modern Medicine*. Montreal: Epimed International, 1995.
13. Neiss ES, Boyd TA. Pharmogenology: The industrial new drug development process, pp. 1–32 in: *The Clinical Research Process in the Pharmaceutical Industry*. Edited by GM Matoren. New York and Basel: Marcel Drekker Inc., 1984.

14. Tannock I, Warr D. Nonrandomized clinical trials of cancer chemotherapy: Phase II or III? *JNCI*,1988;**80**:800–1.
15. The Protocol Review Committee, The Data Center, The Research and Treatment Division and The New Drug Development Office. Phase II trials in the EORTC. *Eur J Cancer*,1997;**33**:1361–3.
16. Shoemaker D, Burke G, Dorr A, Temple R, Friedman MA. A regulatory perspective, pp. 193–201 in: *Quality of Life Assessments in Clinical Trials.* Edited by B. Spilker. New York: Raven Press, 1990.
17. Rose G. Sick individuals and sick populations. *Int J Epidemiol*,1985;**14**:32–8.
18. Reiser SJ. Humanism and fact-finding in medicine. *N Engl J Med*,1978;**229**:950–3.
19. Buck C. Popper's philosophy for epidemiologists. *Int J Epidemiol*, 1975;**4**: 159–68.
20. Velanovich V. The function of the case report in medical epistemology. *Theor Surg*, 1992;**7**: 91–4.

A LAYMAN'S GLOSSARY

Absolute risk reduction (ARR): The absolute arithmetical difference in bad event rates for treated and untreated subjects in clinical trials or observational analytical studies. Synonym of *attributable risk* or *risk difference*. May also be expressed as an *absolute benefit increase*.

Case (in general): An individual, a given situation, an occurrence or an event in a particular area of daily or professional life.

Case (in medicine): A particular instance of disease, as in *a case of leukaemia*. Sometimes used incorrectly to designate the patient with the disease. In this book, a patient has a specific case of a given health problem.

Case fatality (rate): The proportion of cases of patients with a specified condition who die of it within a specified time.

Case report (clinical): A structured form of scientific and professional communication normally focused on an unusual single event (patient or clinical situation). The case report aims to provide a better understanding of a case and of its effects on improved clinical decision making. A summary of a case study.

Case study: A detailed description and analysis of an individual case which explains the dynamics, pathology, management and/or outcome of a given disease.

Case series study: A detailed description and analysis of a series of cases which explains the dynamics, pathology, management and/or outcome of a given disease.

In *epidemiological terms,* it refers to the study of several individuals without denominators.

Casuist (in medicine): A practitioner of the study of clinical cases.

Casuistics (in philosophy): The application of general laws and rules to a particular area or fact. A method of solving problems by implementing actual actions based on the general principles and study of similar cases.

Casuistics (in medicine): The recording and study of cases of any disease. The observation, analysis and interpretation of clinical cases. The art of choosing, gathering, structuring and conveying pragmatic information about relevant clinical cases. Casuistics aims to provide a better understanding of a given health problem and therefore to improve clinical decisions.

Casuistry (in philosophy and theology): A system of rules for distinguishing right from wrong in everyday situations, usually associated with a concept of morality that views correct behaviour in terms of obedience to a set of closely defined laws. For some, it implies specious reasoning based on an excessive amount of minute, often unimportant detail. For hospital ethicists, it refers to the art of applying abstract principles, paradigms and analogies to particular cases.

Clinical data: Clinical observations as seen, measured and recorded. Example: blood pressure 120/80mmHg.

Clinical information: The interpretation and meaning given to clinical data. Example: a normotensive patient.

Clinimetrics:	A field focused on indexes, rating scales and other expressions used to describe and measure symptoms, physical signs and other distinctly clinical phenomena in clinical medicine. A process extending from the retrieval of clinical observations to their description, interpretation, classification and categorization.
Collective case study:	See *Case series study*.
Co-morbidity:	All other health problems present in a patient studied and treated for a disease of interest. Example: diabetes or hypertension in a patient studied and treated for cancer.
Conceptual criteria or definitions:	The qualitative characterizations of phenomena under study. Example: a hypertensive patient under study.
Criterion:	A principle or standard by which something is judged. An element that serves as a basis for comparison (standard). Rules according to which observations are selected, measured, classified and interpreted. The term is used in the same way for recruitment of cases.
Cross-sectional study:	A study based on a single observation and evaluation of cases. A *'single shot' study*, based on one examination of patients only. Synonym of *transversal study*.
Deductive research:	Research based on the verification of the acceptability of *a priori* formulated hypotheses. Refers to research or studies that are built specifically to accept or refute a given question.
Evidence:	A fact or body of facts on which a proof, belief or judgement is based. Evidence *does not mean certainty*. Rather, it represents an available proof with varying degrees of certainty.

Evidence-based clinical case reporting:

Critical case reporting based on the strongest evidence taken from the case observation itself and from the related literature.

Evidence table:

In meta-analysis, a *systematic* tabular compilation of the present and absent characteristics of original studies. In clinical case series studies, a similar compilation and review of case, time and place characteristics.

Evidence-based medicine:

The process of systematically finding, assessing and using contemporaneous research results as the basis for clinical decisions. (Always look for the best available information and use it!) The application of simple rules of science and common sense to determine the validity of the information. The application of valid information to answer the clinical question. Patient care based on evidence derived from the best available ('gold standard') studies.

Gradient of the disease:

The directional expression of the severity of a disease, similar to colour shades from light to dark, i.e. from unapparent to fatal cases of a disease.

Hard data:

Clinical and paraclinical data that can be precisely defined and measured. Examples: heart rate, blood cell count.

Hardening of soft data:

All means used to improve the criteria, measurement and quanti-fication of soft data in order for the quality of soft data to match that of hard data as closely as possible.

Hermeneutics:

A philosophical concept of inter-pretation and understanding; making sense of a mess of clinical and paraclinical data and information in a case. Derived from Hermes, the

	messenger of the gods in Greek mythology.
Hypothesis:	A proposal or question that research or study should accept or refute.
Idiographic analysis:	An analysis that is exclusive (or limited) to the case under study.
Idiographic approach (in philosophy):	The advancement of science based on the information or what was acquired as information from one individual case.
Inception moment:	Any moment at which the follow-up of a case begins: beginning of a clinical stage of disease, date of admission, etc. Must always be defined.
Incidence:	The number of *new* events occurring in a given population in a defined period of time. For a more refined definition, see the epidemiological literature.
Index case:	The first case in a family or other defined group (disease sufferers etc.) to come to the attention of the investigator. Often used to formulate preliminary hypotheses and define criteria for selection and follow-up of cases to come, as in the investigation of an infectious disease outbreak. In medical genetics, an index case represents an original patient (propositus or proband) who provides the stimulus to study other members of the family in order to ascertain a possible genetic factor in causation of the presenting condition.
Inductive research:	Research that proceeds from observations serving as a basis for hypotheses and answers. (Hypotheses are a product of the data that precedes them. Studies that originate data are not necessarily built to verify hypotheses and questions of interest.)

Initial state:	The state of the patient (case) before clinical actions of interest (manoeuvres) are applied.
Instrumental case study:	A study focusing on a better understanding of the *problem* represented by the case.
Intrinsic case study:	A study focusing on a better understanding of the *case itself*.
Longitudinal study:	The follow-up of cases in time based on more than one examination. A *'burst of shots' study*. Synonym of *cohort study*.
Manoeuvre:	A clinical action under study, i.e. diagnostic procedure, surgical treatment, medical treatment or care etc.
Meta-analysis:	A systematic, organized and structured evaluation and synthesis of a problem of interest based on the results of many independent studies of that problem (disease cause, occurrence, treatment effect, diagnostic method, prognosis etc.). Epidemiology of the results of independent studies of the same problem of interest. A study of studies. A similar precise integration of cases in case series studies becomes a *meta-analysis of cases*.
Monographic study:	A study as detailed and complete as possible.
n-of-1 research or trial:	Any research on a single patient. In *clinical epidemiology*, it is a variation of a randomized controlled trial in which a sequence of alternative treatment regimens is randomly assigned to a patient.

Nomothetic approach (in philosophy):	The advancement of science based on the information or on what was acquired as information from a set of cases.
Number needed to treat (NNT):	The number of patients who need to be treated to achieve one additional favourable outcome. It is calculated as 1/absolute risk reduction.
Occurrence study:	Any study of frequency of disease or other attribute or event in a population. Usually a descriptive observational study of the prevalence, incidence, mortality or case fatality of a disease or of the prevailing characteristics of the individuals, time or place where a phenomenon of interest 'occurs'.
Operational criteria:	Measurable rules of selection of cases and variables. Example: the cases to follow will be subjects with a blood pressure of 160/120mmHg or higher than either of these values.
Outcome:	Any possible result that may stem from exposure to a causal factor, or from preventive and therapeutic interventions. All identified changes in health status arising as a consequence of the handling of a health problem.
Paradigm:	Any pattern or example. The way we look at things, events and actions around us.
Phase I clinical trial:	Clinical trial whose objective is to determine how *healthy individuals* will respond to the treatment (pharmacodynamics, tolerance, metabolism, adverse effects).
Phase II clinical trial:	Clinical trials whose objective is to determine how the disease *sufferer* will respond to treatment. Either one group of patients (early Phase II) or two or more groups (without *a priori*

	defined statistical considerations) are studied (late Phase II).
Phase III clinical trial:	A randomized controlled clinical trial. Subjects are randomly divided into at least two groups to be compared. Several factors are not revealed to the patients and investigators, such as the group to which the patients belong and the true outcome measured.
Phase IV clinical trial:	Similar to Phase III but without preselected patients. The patients enter the trial 'as they come through the door'. 'Clean' or 'neat' cases participate.
Phase V clinical trial:	In this situation, non-selected patients who suffer from various co-morbidities (diseases other than the one under study) and are treated for these additional health problems (co-treatments are present) participate in post-marketing studies.
Phenomenology:	The philosophical concept of research on the meaning of the experience lived by a case.
Prevalence:	Cases existing at a given moment of study or observation. A more refined concept of prevalence is available in the epidemiological literature.
Prognosis:	An assessment of the patient's future, based on probabilistic considerations of various beneficial and detrimental clinical outcomes as causally or otherwise determined by various clinical factors and biological and social characteristics of the patient and of the pathology (disease source) itself.
Prognostic characteristics:	Characteristics of patients, a time or a place related to a particular probability of events in individuals

	who already have the disease under study. A common term for prognostic factors and prognostic markers (see below).
Prognostic factors:	*Modifiable* patient, time or place characteristics related to the outcome of the health problem under study. Example: antibiotic treatment for infection.
Prognostic markers:	*Non modifiable* patient, time or place characteristics related to the outcome of the health problem under study. Example: age in degenerative joint diseases.
Qualitative meta-analysis:	A systematic overview of characteristics and components of original studies of the same problem. A method of assessing the importance and relevance of medical information through a general, systematic and uniform application of pre-established criteria of acceptability of original studies representing the body of knowledge of a given health problem or question. A systematic assessment of the completeness and quality of the characteristics of cases represents a *qualitative meta-analysis of cases*.
Qualitative research:	Any kind of research that produces findings not arrived at by means of statistical procedures or other means of quantification. Some data may be quantified but the analysis itself is a qualitative one. An in-depth study of unique observations.
Quantitative meta-analysis:	The general, systematic and uniform evaluation and integration of *dimensions*, i.e. numerical findings, from independent studies of the same problem of interest. 'Typical' values for sets of studies are of primary interest. Examples: typical

odds ratios, protective efficacy rates etc.

Quantitative research: Research based on series of observations, where phenomena are measured, quantified, counted, described and analysed by statistical methods. An in-depth study of multiple observations.

Rate: An expression of the frequency at which an event occurs in a defined population. The number of events (numerator) is related to all individuals that take part in them (denominator). More refined definitions are available in the epidemiological literature.

Relative risk reduction (RRR): The proportional reduction in rates of bad events between control and experimental participants in a trial. Synonym of *aetiologic fraction, attributable fraction, attributable risk per cent, protective efficacy rate*. May also be expressed as a *relative benefit increase*.

Risk: The probability that an event (disease, complication, improvement etc.) will occur.

Risk characteristics: A common term for risk factors and risk markers (see below).

Risk factor: Any *modifiable* characteristic of persons, times or places related to disease occurrence. Example: smoking as a trigger of lung cancer.

Risk marker: Any *non-modifiable* characteristic of persons, times or places related to disease occurrence. Example: age in relation to cancer or cardiovascular diseases.

S.O.A.P.: Often, the first abbreviation that is learnt by any house officer writing patients' daily progress reports:

	Subjective perception of the patient, **O**bjective data, **A**ssessment, and **P**lan of management of the case.
Soft data:	Clinical and paraclinical observations that are difficult to define, measure and classify. Examples: sorrow, anxiety, paraesthesia.
Spectrum of the disease:	The variability of the clinical picture in terms of the extent of the disease; similar to a spectrum or rainbow of colours. Example: all the systemic manifestations of infectious diseases.
Subsequent state:	The state of the patient (case) following the clinical management (actions, manoeuvres) under study. It may or may not be the result of this management.
Theory:	An interrelated set of constructs (variables) formed into propositions or hypotheses that specify the relationship among variables.

REFERENCES

1. *Dorland's Illustrated Medical Dictionary*, 24th Edition. Philadelphia and London: WB Saunders, 1965.
2. *Miller-Keane Encyclopedia & Dictionary of Medicine, Nursing & Allied Health*, 6th Edition. Philadelphia: WB Saunders, 1997.
3. *A Dictionary of Epidemiology*, 3rd Edition. Edited by JM Last. New York, Oxford and Toronto: Oxford University Press, 1995.
4. *New Illustrated Webster's Dictionary Including Thesaurus of Synonyms & Antonyms.* New York: Pamco Publishing Company, Inc., 1992.
5. Jenicek M. *Epidemiology. The Logic of Modern Medicine.* Montreal: Epimed International, 1995.
6. Related references quoted in the text of this book.

INDEX

Page numbers printed in **bold** type refer to figures and tables; those in *italic* relate to entries in the glossary.

Fallopian tube carcinoma, case series study, 113–14
Floor presentations, 6

Generalizations, 24, 67–8
  v. individual observations/cases, 132
Gradient of the disease, 59, 60, *138*

Hallucinations, 104, **105**
Hard data, 49, *138*
Hardening of soft data, 49, *138*
Hermeneutics, 54, *138–9*
Hospital-focused case presentations, 43–4
Hypotheses, 33, 128, *139, 139–40*

Idiographic analysis, 67, *139*
Idiographic approach, 19, *139*
Immaculate Conception, 29, 118–19
Immunizations, case series study, 107
Inception moment, *139*
Incidence, *139*
Incidence studies, **28**
Index case(s), 4, 27, 101
  definition, 27, *139*
  examples, 27–9
Induction period, of disease, 59
Inductive research, 22, 24, *139–40*
Initial state, 55, 58, *140*
  in case report example, 89
Instrumental case studies, 24, *140*
Intensive designs, 19
Interpretive meta-synthesis, 113

Intra-subject-replication designs, 19
Intrinsic case studies, 23, *140*

Journal of Obstetrics and Gynecology, 7
Journal of the American Medical Association (JAMA), 51, 72–4
Journals *see* Medical journals
Justification, 54
  in case report example, 92, 94

The Lancet, 7, 45, 51
Longitudinal studies, 102, *140*
  see also Cohort studies

Manoeuvres, 56, *140*
Medical casuistics, 16, 17–18, 27–9, 30–2
  bibliography, 42–3
  challenges of, 130–1
  method of handling cases, 131–2
  as a science, 132
Medical history, in case report example, 87–8
Medical investigation, steps in, 29
Medical journals
  case reporting in, 6, 7
  guidelines to contributors, 42, 44
  journals' expectations of case reports, 43, 74–5
  see also Publishing
Meta-analysis, 21, 26
  of cases, 102, 111–15, *140*
  definition, 112, *140*
  qualitative, 114–15, *143*
  quantitative, 114, *143*

Training
  and clinical case studies, 30–1
  in study of individuals, 6
Transversal studies *see* Cross-
    sectional studies
Traumatic optic neuropathy,
  case series study, 114

Treatment, 61–2, 68–9
  in case report example, 89–91
Typical odds ratios, 114

Ward case reports, routine, 43–4
Ward presentations, 6